Liquid Pipeline Hydraulics

MECHANICAL ENGINEERING
A Series of Textbooks and Reference Books

Founding Editor

L. L. Faulkner

Columbus Division, Battelle Memorial Institute
and Department of Mechanical Engineering
The Ohio State University
Columbus, Ohio

1. *Spring Designer's Handbook*, Harold Carlson
2. *Computer-Aided Graphics and Design*, Daniel L. Ryan
3. *Lubrication Fundamentals*, J. George Wills
4. *Solar Engineering for Domestic Buildings*, William A. Himmelman
5. *Applied Engineering Mechanics: Statics and Dynamics*, G. Boothroyd and C. Poli
6. *Centrifugal Pump Clinic*, Igor J. Karassik
7. *Computer-Aided Kinetics for Machine Design*, Daniel L. Ryan
8. *Plastics Products Design Handbook, Part A: Materials and Components; Part B: Processes and Design for Processes*, edited by Edward Miller
9. *Turbomachinery: Basic Theory and Applications*, Earl Logan, Jr.
10. *Vibrations of Shells and Plates*, Werner Soedel
11. *Flat and Corrugated Diaphragm Design Handbook*, Mario Di Giovanni
12. *Practical Stress Analysis in Engineering Design*, Alexander Blake
13. *An Introduction to the Design and Behavior of Bolted Joints*, John H. Bickford
14. *Optimal Engineering Design: Principles and Applications*, James N. Siddall
15. *Spring Manufacturing Handbook*, Harold Carlson
16. *Industrial Noise Control: Fundamentals and Applications*, edited by Lewis H. Bell
17. *Gears and Their Vibration: A Basic Approach to Understanding Gear Noise*, J. Derek Smith
18. *Chains for Power Transmission and Material Handling: Design and Applications Handbook*, American Chain Association
19. *Corrosion and Corrosion Protection Handbook*, edited by Philip A. Schweitzer
20. *Gear Drive Systems: Design and Application*, Peter Lynwander
21. *Controlling In-Plant Airborne Contaminants: Systems Design and Calculations*, John D. Constance
22. *CAD/CAM Systems Planning and Implementation*, Charles S. Knox
23. *Probabilistic Engineering Design: Principles and Applications*, James N. Siddall
24. *Traction Drives: Selection and Application*, Frederick W. Heilich III and Eugene E. Shube
25. *Finite Element Methods: An Introduction*, Ronald L. Huston and Chris E. Passerello

26. *Mechanical Fastening of Plastics: An Engineering Handbook*, Brayton Lincoln, Kenneth J. Gomes, and James F. Braden
27. *Lubrication in Practice: Second Edition*, edited by W. S. Robertson
28. *Principles of Automated Drafting*, Daniel L. Ryan
29. *Practical Seal Design*, edited by Leonard J. Martini
30. *Engineering Documentation for CAD/CAM Applications*, Charles S. Knox
31. *Design Dimensioning with Computer Graphics Applications*, Jerome C. Lange
32. *Mechanism Analysis: Simplified Graphical and Analytical Techniques*, Lyndon O. Barton
33. *CAD/CAM Systems: Justification, Implementation, Productivity Measurement*, Edward J. Preston, George W. Crawford, and Mark E. Coticchia
34. *Steam Plant Calculations Manual*, V. Ganapathy
35. *Design Assurance for Engineers and Managers*, John A. Burgess
36. *Heat Transfer Fluids and Systems for Process and Energy Applications*, Jasbir Singh
37. *Potential Flows: Computer Graphic Solutions*, Robert H. Kirchhoff
38. *Computer-Aided Graphics and Design: Second Edition*, Daniel L. Ryan
39. *Electronically Controlled Proportional Valves: Selection and Application*, Michael J. Tonyan, edited by Tobi Goldoftas
40. *Pressure Gauge Handbook*, AMETEK, U.S. Gauge Division, edited by Philip W. Harland
41. *Fabric Filtration for Combustion Sources: Fundamentals and Basic Technology*, R. P. Donovan
42. *Design of Mechanical Joints*, Alexander Blake
43. *CAD/CAM Dictionary*, Edward J. Preston, George W. Crawford, and Mark E. Coticchia
44. *Machinery Adhesives for Locking, Retaining, and Sealing*, Girard S. Haviland
45. *Couplings and Joints: Design, Selection, and Application*, Jon R. Mancuso
46. *Shaft Alignment Handbook*, John Piotrowski
47. *BASIC Programs for Steam Plant Engineers: Boilers, Combustion, Fluid Flow, and Heat Transfer*, V. Ganapathy
48. *Solving Mechanical Design Problems with Computer Graphics*, Jerome C. Lange
49. *Plastics Gearing: Selection and Application*, Clifford E. Adams
50. *Clutches and Brakes: Design and Selection*, William C. Orthwein
51. *Transducers in Mechanical and Electronic Design*, Harry L. Trietley
52. *Metallurgical Applications of Shock-Wave and High-Strain-Rate Phenomena*, edited by Lawrence E. Murr, Karl P. Staudhammer, and Marc A. Meyers
53. *Magnesium Products Design*, Robert S. Busk
54. *How to Integrate CAD/CAM Systems: Management and Technology*, William D. Engelke
55. *Cam Design and Manufacture: Second Edition*; with cam design software for the IBM PC and compatibles, disk included, Preben W. Jensen
56. *Solid-State AC Motor Controls: Selection and Application*, Sylvester Campbell
57. *Fundamentals of Robotics*, David D. Ardayfio
58. *Belt Selection and Application for Engineers*, edited by Wallace D. Erickson
59. *Developing Three-Dimensional CAD Software with the IBM PC*, C. Stan Wei
60. *Organizing Data for CIM Applications*, Charles S. Knox, with contributions by Thomas C. Boos, Ross S. Culverhouse, and Paul F. Muchnicki

61. *Computer-Aided Simulation in Railway Dynamics*, by Rao V. Dukkipati and Joseph R. Amyot
62. *Fiber-Reinforced Composites: Materials, Manufacturing, and Design*, P. K. Mallick
63. *Photoelectric Sensors and Controls: Selection and Application*, Scott M. Juds
64. *Finite Element Analysis with Personal Computers*, Edward R. Champion, Jr., and J. Michael Ensminger
65. *Ultrasonics: Fundamentals, Technology, Applications: Second Edition, Revised and Expanded*, Dale Ensminger
66. *Applied Finite Element Modeling: Practical Problem Solving for Engineers*, Jeffrey M. Steele
67. *Measurement and Instrumentation in Engineering: Principles and Basic Laboratory Experiments*, Francis S. Tse and Ivan E. Morse
68. *Centrifugal Pump Clinic: Second Edition, Revised and Expanded*, Igor J. Karassik
69. *Practical Stress Analysis in Engineering Design: Second Edition, Revised and Expanded*, Alexander Blake
70. *An Introduction to the Design and Behavior of Bolted Joints: Second Edition, Revised and Expanded*, John H. Bickford
71. *High Vacuum Technology: A Practical Guide*, Marsbed H. Hablanian
72. *Pressure Sensors: Selection and Application*, Duane Tandeske
73. *Zinc Handbook: Properties, Processing, and Use in Design*, Frank Porter
74. *Thermal Fatigue of Metals*, Andrzej Weronski and Tadeusz Hejwowski
75. *Classical and Modern Mechanisms for Engineers and Inventors*, Preben W. Jensen
76. *Handbook of Electronic Package Design*, edited by Michael Pecht
77. *Shock-Wave and High-Strain-Rate Phenomena in Materials*, edited by Marc A. Meyers, Lawrence E. Murr, and Karl P. Staudhammer
78. *Industrial Refrigeration: Principles, Design and Applications*, P. C. Koelet
79. *Applied Combustion*, Eugene L. Keating
80. *Engine Oils and Automotive Lubrication*, edited by Wilfried J. Bartz
81. *Mechanism Analysis: Simplified and Graphical Techniques, Second Edition, Revised and Expanded*, Lyndon O. Barton
82. *Fundamental Fluid Mechanics for the Practicing Engineer*, James W. Murdock
83. *Fiber-Reinforced Composites: Materials, Manufacturing, and Design, Second Edition, Revised and Expanded*, P. K. Mallick
84. *Numerical Methods for Engineering Applications*, Edward R. Champion, Jr.
85. *Turbomachinery: Basic Theory and Applications, Second Edition, Revised and Expanded*, Earl Logan, Jr.
86. *Vibrations of Shells and Plates: Second Edition, Revised and Expanded*, Werner Soedel
87. *Steam Plant Calculations Manual: Second Edition, Revised and Expanded*, V. Ganapathy
88. *Industrial Noise Control: Fundamentals and Applications, Second Edition, Revised and Expanded*, Lewis H. Bell and Douglas H. Bell
89. *Finite Elements: Their Design and Performance*, Richard H. MacNeal
90. *Mechanical Properties of Polymers and Composites: Second Edition, Revised and Expanded*, Lawrence E. Nielsen and Robert F. Landel
91. *Mechanical Wear Prediction and Prevention*, Raymond G. Bayer

92. *Mechanical Power Transmission Components*, edited by David W. South and Jon R. Mancuso
93. *Handbook of Turbomachinery*, edited by Earl Logan, Jr.
94. *Engineering Documentation Control Practices and Procedures*, Ray E. Monahan
95. *Refractory Linings Thermomechanical Design and Applications*, Charles A. Schacht
96. *Geometric Dimensioning and Tolerancing: Applications and Techniques for Use in Design, Manufacturing, and Inspection*, James D. Meadows
97. *An Introduction to the Design and Behavior of Bolted Joints: Third Edition, Revised and Expanded*, John H. Bickford
98. *Shaft Alignment Handbook: Second Edition, Revised and Expanded*, John Piotrowski
99. *Computer-Aided Design of Polymer-Matrix Composite Structures*, edited by Suong Van Hoa
100. *Friction Science and Technology*, Peter J. Blau
101. *Introduction to Plastics and Composites: Mechanical Properties and Engineering Applications*, Edward Miller
102. *Practical Fracture Mechanics in Design*, Alexander Blake
103. *Pump Characteristics and Applications*, Michael W. Volk
104. *Optical Principles and Technology for Engineers*, James E. Stewart
105. *Optimizing the Shape of Mechanical Elements and Structures*, A. A. Seireg and Jorge Rodriguez
106. *Kinematics and Dynamics of Machinery*, Vladimír Stejskal and Michael Valášek
107. *Shaft Seals for Dynamic Applications*, Les Horve
108. *Reliability-Based Mechanical Design*, edited by Thomas A. Cruse
109. *Mechanical Fastening, Joining, and Assembly*, James A. Speck
110. *Turbomachinery Fluid Dynamics and Heat Transfer*, edited by Chunill Hah
111. *High-Vacuum Technology: A Practical Guide, Second Edition, Revised and Expanded*, Marsbed H. Hablanian
112. *Geometric Dimensioning and Tolerancing: Workbook and Answerbook*, James D. Meadows
113. *Handbook of Materials Selection for Engineering Applications*, edited by G. T. Murray
114. *Handbook of Thermoplastic Piping System Design*, Thomas Sixsmith and Reinhard Hanselka
115. *Practical Guide to Finite Elements: A Solid Mechanics Approach*, Steven M. Lepi
116. *Applied Computational Fluid Dynamics*, edited by Vijay K. Garg
117. *Fluid Sealing Technology*, Heinz K. Muller and Bernard S. Nau
118. *Friction and Lubrication in Mechanical Design*, A. A. Seireg
119. *Influence Functions and Matrices*, Yuri A. Melnikov
120. *Mechanical Analysis of Electronic Packaging Systems*, Stephen A. McKeown
121. *Couplings and Joints: Design, Selection, and Application, Second Edition, Revised and Expanded*, Jon R. Mancuso
122. *Thermodynamics: Processes and Applications*, Earl Logan, Jr.
123. *Gear Noise and Vibration*, J. Derek Smith
124. *Practical Fluid Mechanics for Engineering Applications*, John J. Bloomer
125. *Handbook of Hydraulic Fluid Technology*, edited by George E. Totten
126. *Heat Exchanger Design Handbook*, T. Kuppan

127. *Designing for Product Sound Quality*, Richard H. Lyon
128. *Probability Applications in Mechanical Design*, Franklin E. Fisher and Joy R. Fisher
129. *Nickel Alloys*, edited by Ulrich Heubner
130. *Rotating Machinery Vibration: Problem Analysis and Troubleshooting*, Maurice L. Adams, Jr.
131. *Formulas for Dynamic Analysis*, Ronald L. Huston and C. Q. Liu
132. *Handbook of Machinery Dynamics*, Lynn L. Faulkner and Earl Logan, Jr.
133. *Rapid Prototyping Technology: Selection and Application*, Kenneth G. Cooper
134. *Reciprocating Machinery Dynamics: Design and Analysis*, Abdulla S. Rangwala
135. *Maintenance Excellence: Optimizing Equipment Life-Cycle Decisions*, edited by John D. Campbell and Andrew K. S. Jardine
136. *Practical Guide to Industrial Boiler Systems*, Ralph L. Vandagriff
137. *Lubrication Fundamentals: Second Edition, Revised and Expanded*, D. M. Pirro and A. A. Wessol
138. *Mechanical Life Cycle Handbook: Good Environmental Design and Manufacturing*, edited by Mahendra S. Hundal
139. *Micromachining of Engineering Materials*, edited by Joseph McGeough
140. *Control Strategies for Dynamic Systems: Design and Implementation*, John H. Lumkes, Jr.
141. *Practical Guide to Pressure Vessel Manufacturing*, Sunil Pullarcot
142. *Nondestructive Evaluation: Theory, Techniques, and Applications*, edited by Peter J. Shull
143. *Diesel Engine Engineering: Thermodynamics, Dynamics, Design, and Control*, Andrei Makartchouk
144. *Handbook of Machine Tool Analysis*, Ioan D. Marinescu, Constantin Ispas, and Dan Boboc
145. *Implementing Concurrent Engineering in Small Companies*, Susan Carlson Skalak
146. *Practical Guide to the Packaging of Electronics: Thermal and Mechanical Design and Analysis*, Ali Jamnia
147. *Bearing Design in Machinery: Engineering Tribology and Lubrication*, Avraham Harnoy
148. *Mechanical Reliability Improvement: Probability and Statistics for Experimental Testing*, R. E. Little
149. *Industrial Boilers and Heat Recovery Steam Generators: Design, Applications, and Calculations*, V. Ganapathy
150. *The CAD Guidebook: A Basic Manual for Understanding and Improving Computer-Aided Design*, Stephen J. Schoonmaker
151. *Industrial Noise Control and Acoustics*, Randall F. Barron
152. *Mechanical Properties of Engineered Materials*, Wolé Soboyejo
153. *Reliability Verification, Testing, and Analysis in Engineering Design*, Gary S. Wasserman
154. *Fundamental Mechanics of Fluids: Third Edition*, I. G. Currie
155. *Intermediate Heat Transfer*, Kau-Fui Vincent Wong
156. *HVAC Water Chillers and Cooling Towers: Fundamentals, Application, and Operation*, Herbert W. Stanford III
157. *Gear Noise and Vibration: Second Edition, Revised and Expanded*, J. Derek Smith

158. *Handbook of Turbomachinery: Second Edition, Revised and Expanded,* edited by Earl Logan, Jr., and Ramendra Roy
159. *Piping and Pipeline Engineering: Design, Construction, Maintenance, Integrity, and Repair,* George A. Antaki
160. *Turbomachinery: Design and Theory,* Rama S. R. Gorla and Aijaz Ahmed Khan
161. *Target Costing: Market-Driven Product Design,* M. Bradford Clifton, Henry M. B. Bird, Robert E. Albano, and Wesley P. Townsend
162. *Fluidized Bed Combustion,* Simeon N. Oka
163. *Theory of Dimensioning: An Introduction to Parameterizing Geometric Models,* Vijay Srinivasan
164. *Handbook of Mechanical Alloy Design*, edited by George E. Totten, Lin Xie, and Kiyoshi Funatani
165. *Structural Analysis of Polymeric Composite Materials,* Mark E. Tuttle
166. *Modeling and Simulation for Material Selection and Mechanical Design,* edited by George E. Totten, Lin Xie, and Kiyoshi Funatani
167. *Handbook of Pneumatic Conveying Engineering,* David Mills, Mark G. Jones, and Vijay K. Agarwal
168. *Clutches and Brakes: Design and Selection, Second Edition*, William C. Orthwein
169. *Fundamentals of Fluid Film Lubrication: Second Edition,* Bernard J. Hamrock, Steven R. Schmid, and Bo O. Jacobson
170. *Handbook of Lead-Free Solder Technology for Microelectronic Assemblies*, edited by Karl J. Puttlitz and Kathleen A. Stalter
171. *Vehicle Stability*, Dean Karnopp
172. *Mechanical Wear Fundamentals and Testing: Second Edition, Revised and Expanded,* Raymond G. Bayer
173. *Liquid Pipeline Hydraulics,* E. Shashi Menon
174. *Solid Fuels Combustion and Gasification*, Marcio L. de Souza-Santos
175. *Mechanical Tolerance Stackup and Analysis*, Bryan R. Fischer
176. *Engineering Design for Wear*, Raymond G. Bayer
177. *Vibrations of Shells and Plates: Third Edition, Revised and Expanded,* Werner Soedel
178. *Refractories Handbook*, edited by Charles A. Schacht
179. *Practical Engineering Failure Analysis*, Hani Tawancy, Halim Hamid, Nureddin M. Abbas
180. *Mechanical Alloying and Milling*, C. Suryanarayana

Additional Volumes in Preparation

Design of Automatic Machinery, Stephen J. Derby

Mechanical Vibration: Analysis, Uncertainties, and Control, Second Edition, Revised and Expanded, Haym Benaroya

Practical Fracture Mechanics in Design: Second Edition, Revised and Expanded, Arun Shukla

Mechanical Engineering Software

Spring Design with an IBM PC, Al Dietrich

Mechanical Design Failure Analysis: With Failure Analysis System Software for the IBM PC, David G. Ullman

Liquid Pipeline Hydraulics

E. Shashi Menon
SYSTEK Technologies, Inc.
Lake Havasu City, Arizona, U.S.A.

CRC Press
Taylor & Francis Group
Boca Raton London New York

CRC Press is an imprint of the
Taylor & Francis Group, an **informa** business

CRC Press
Taylor & Francis Group
6000 Broken Sound Parkway NW, Suite 300
Boca Raton, FL 33487-2742

First issued in paperback 2019

ISBN-13: 978-0-8247-5317-7 (hbk)
ISBN-13: 978-0-367-39415-8 (pbk)

Library of Congress Cataloging-in-Publication Data
A catalog record for this book is available from the Library of Congress.

Visit the Taylor & Francis Web site at
http://www.taylorandfrancis.com

and the CRC Press Web site at
http://www.crcpress.com

Preface

This book presents liquid pipeline hydraulics as it applies to transportation of liquids through pipelines in a single-phase steady state environment. It serves as a practical handbook for engineers, technicians, and others involved in design and operation of pipelines transporting liquids. Existing books on the subject are mathematically rigorous and theoretical but lack practical applications. Using this book, engineers can better understand and apply the principles of hydraulics to their daily work in the pipeline industry without resorting to complicated formulas and theorems. Numerous examples from my experience are included to illustrate application of pipeline hydraulics.

The application of hydraulics to liquid pipelines involves an understanding of various properties of liquids, the concept of pressure, friction, and calculation of the energy required to transport liquid from point A to point B through a pipeline. You will not find rigorous mathematical derivation of formulas in this book. The formulas necessary for calculations are presented and described without using calculus or complex mathematical methods. The reader interested in how the formulas and equations are derived should refer to the books and other publications in the References section.

This book covers liquid properties that affect flow through pipelines, calculation of pressure drop due to friction, horsepower, and the number of pump stations required for transporting a liquid through a pipeline. Among the topics considered are the basic equations necessary for pipeline design, commonly used formulas to calculate frictional pressure drop and necessary horsepower, the feasibility of improving an existing pipeline's performance using drag-reduction additives, and power optimization studies. The use of pumps and valves in pipelines is addressed along with modifications necessary to improve pipeline throughput. Economic analysis and transportation tariff calculations are also included.

The book can be used for the analysis of liquid pipeline gathering systems, plant or terminal piping, and long-distance trunk lines. The primary audience is engineers and technicians working in the petroleum, water, and process industries. It could also be used as a text for a college-level course in liquid pipeline hydraulics.

E. Shashi Menon

Contents

Liquid Pipeline Hydraulics

1

Introduction

Pipeline hydraulics deals with the flow of fluids in pipelines. Fluids are defined as liquids and gases. Specifically, this book deals with liquid flow in pipelines. Liquids are considered to be incompressible for most purposes. Today, thousands of miles of pipelines are used to transport crude oil and petroleum products such as gasoline and diesel from refineries to storage tanks and delivery terminals. Similarly, thousands of miles of concrete and steel pipelines are used to transport water from reservoirs to distribution locations. Engineers are interested in the effect of pipe size, liquid properties, pipe length, etc., in determining the pressure required and horsepower necessary for transporting a liquid from point A to point B in a pipeline. It is clear that the heavier the liquid, the more pressure and hence more horsepower required to transport a given quantity for a specified distance. In all these cases we are interested in determining the optimum pipe size required to transport given volumes of liquids economically and safely through the pipelines.

This book consists of 12 chapters that cover the practical aspects of liquid pipeline hydraulics and the economics of pipelines used to transport liquids under steady-state conditions, except Chapter 11, which introduces the reader to unsteady flow. For a more detailed analysis and study of unsteady flow and transients, the reader should consult one of the books listed in the References section at the end of the book. Appendices containing tables and charts, answers to selected problems, and a summary of formulas are also included at the end of this book.

Chapter 2 covers units of measurement, and properties of liquids such as density, gravity, and viscosity that are important in liquid pipeline hydraulics. Chapter 3 discusses pressure, velocity, Reynolds number, friction factor, and pressure drop calculations using various formulas. Several example problems are discussed and solved to illustrate the various methods currently used in pipeline engineering.

Chapter 4 is devoted to the strength analysis of pipes. It addresses allowable internal working pressures and hydrostatic test pressures and how they are calculated.

Chapter 5 extends the concepts developed in Chapter 3 by analyzing the total pressure and horsepower required to pump a liquid through long-distance pipelines with multiple pump stations, including the transportation of high vapor pressure liquids, such as liquefied petroleum gas (LPG). Injection and delivery along a long pipeline and branch pipe analysis are also covered, and the use of pipe loops to reduce friction and increase throughput is analyzed.

Chapter 6 deals with optimizing pump station locations in a trunk line, minimizing pipe wall thickness using telescoping and pipe grade tapering. Open channel flow, slack line operation in hilly terrain, and batching different products are also addressed in this chapter.

Chapter 7 covers centrifugal pumps and positive displacement pumps applied to pipeline transportation. Centrifugal pump performance curves, Affinity Laws, and the effect of viscosity are discussed as well as the importance of net positive suction head (NPSH). Operation of pumps in series and parallel, and modifications needed to operate pumps effectively, are also covered in this chapter.

Chapter 8 discusses pump station design, minimizing energy loss due to pump throttling with constant-speed motor-driven pumps. The advantages of using variable speed drive (VSD) pumps are also explained and illustrated with examples.

Chapter 9 introduces the reader to thermal hydraulics, pressure drop calculations, and temperature profiles in a buried heated liquid pipeline. The importance of thermal conductivity, overall heat transfer coefficient, and how they affect heat loss to the surrounding are covered.

Chapter 10 introduces flow measurement devices used in measuring liquid flow rate in pipelines. Several of the more common instruments such as the Venturi meter, flow nozzle, and orifice meter are discussed and calculation methods explained.

Chapter 11 gives a basic introduction to unsteady flow and transient hydraulic analysis. This is an advanced concept that would require a separate book to cover the subject fully. Therefore, this chapter will serve as a starting point in understanding transient pipeline hydraulics. The reader should consult one of the publications listed in the References section for a more detailed study of unsteady flow and pipeline transients.

Chapter 12 addresses economic aspects related to pipeline feasibility studies. In addition, the pipeline and pump station capital cost, annual operating cost, and calculation of transportation tariff are discussed. Also covered in this chapter is the analysis of the optimum pipe size and pumping equipment required that produces the least cost. A discounted cash flow approach using the present value (PV) of investment is employed in determining the optimum pipe size for a particular application.

In each chapter, example problems are used to illustrate the concepts introduced. Problems for practice are also included at the end of each chapter. Answers to selected problems may be found in Appendix B.

Appendix A consists of tables and charts containing units and conversions, common properties of petroleum fluids, etc.

In addition, for quick reference, formulas used in all chapters have been assembled and summarized in Appendix C.

2

Properties of Liquids

In this chapter we discuss the various units of measurement employed in liquid pipeline hydraulics and proceed to cover the more important properties of liquids that affect hydraulic calculations. The importance of specific gravity, viscosity of pure liquids and mixtures will be analyzed and the concepts will be illustrated with sample problems. This chapter forms the foundation for all calculations involving pipeline pressure drops and horsepower requirements in subsequent chapters. In Appendix A you will find tables listing properties of commonly used liquids such as water and petroleum products.

2.1 Units of Measurement

Before we discuss liquid properties it would be appropriate to identify the different units of measurement used in pipeline hydraulics calculations. Over the years the English-speaking world adopted so-called English units of measurement, while most other European and Asian countries adopted the metric system of units.

The English system of units (referred to in the United States as Customary U.S. units) derives from the old foot-pound-second (FPS) and foot-slug-second (FSS) systems that originated in England. The basic units are foot for length, slug for mass, and second for measurement of time. In the past, the FPS system used pound for mass. Since force, a derived unit, was also measured in pounds, there was evidently some confusion. To clarify, the terms pound-mass (lbm) and pound-force (lbf) were introduced. Numerically, the weight (which is a force due to gravity) of 1 pound-mass was equal to 1 pound-force. However, the introduction of slug for the unit of mass resulted in the adoption of pound exclusively for the unit of force. Thus, in the FSS system which is now used in the United States, the unit of mass is slug. The relationship between a slug, lbf, and lbm will be explained later in this chapter.

In the metric system, originally known as the centimeter-gram-second (CGS) system, the corresponding units for length, mass, and time were centimeter, gram, and second respectively. In later years, a modified metric system called the meter-kilogram-second (MKS) system emerged. In MKS units, the meter was used for the measurement of length and kilogram for the measurement of mass. The measurement for time remained the second for all systems of units.

The scientific and engineering communities have attempted during the last four decades to standardize on a universal system of units worldwide. Through the International Standards Organization (ISO), a policy for an International System of Units (SI) was formulated. The SI units are also known as Système Internationale units.

The conversion from the older system of units to SI units has advanced at different rates in different countries. Most countries of Western Europe and all of Eastern Europe, Russia, India, China, Japan, Australia, and South America have adopted SI units completely. In North America, Canada and Mexico have adopted SI units almost completely. However, engineers and scientists in these countries use both SI units and English units due to their business dealings with the United States. In the United States, SI units are used increasingly in colleges and the scientific community. However, the majority of work is still done using the English units referred to sometimes as Customary U.S. units.

The Metric Conversion Act of 1975 accelerated the adoption of the SI system of units in the United States. The American Society of Mechanical Engineers (ASME), the American Society of Civil Engineers (ASCE), and other professional societies and organizations have assisted in the process of conversion from English to SI units using the respective Institutions' publications. For example, ASME through the ASME Metric Study Committee published a series of articles in the *Mechanical Engineering* magazine to help engineers master the SI system of units.

In the United States, the complete changeover to SI units has not materialized fast enough. Therefore in this transition phase, engineering students, practicing engineers, technicians, and scientists must be familiar with the different systems of units such as English, metric CGS, metric MKS, and SI. In this book we will use both English units (Customary U.S.) and the SI system of units.

Units of measurement are generally divided into three classes as follows:

Base units
Supplementary units
Derived units

By definition, base units are dimensionally independent. These are units of length, mass, time, electric current, temperature, amount of substance, and luminous intensity.

Supplementary units are those used to measure plain angles and solid angles. Examples include the radian and steradian.

Derived units are those that are formed by combination of base units, supplementary units, and other derived units. Examples of derived units are those of force, pressure, and energy.

2.1.1 Base Units

In the English (Customary U.S.) system of units, the following base units are used:

Length	foot (ft)
Mass	slug (slug)
Time	second (s)
Electric current	ampere (A)
Temperature	degree Fahrenheit (°F)
Amount of substance	mole (mol)
Luminous intensity	candela (cd)

In SI units, the following base units are defined:

Length	meter (m)
Mass	kilogram (kg)
Time	second (s)
Electric current	ampere (A)
Temperature	kelvin (K)
Amount of substance	mole (mol)
Luminous intensity	candela (cd)

2.1.2 Supplementary Units

Supplementary units in both the English and SI systems of units are:

Plane angle	radian (rad)
Solid angle	steradian (sr)

The radian is defined as the plane angle between two radii of a circle with an arc length equal to the radius. Thus, it represents the angle of a sector of a circle with the arc length the same as its radius.

The steradian is the solid angle having its apex at the center of a sphere such that the area of the surface of the sphere that it cuts out is equal to that of a square with sides equal to the radius of this sphere.

2.1.3 Derived Units

Derived units are generated from a combination of base units, supplementary units, and other derived units. Examples of derived units include those of area, volume, etc.
In English units the following derived units are used:

Area	square inch (in.^{2}), square foot (ft^2)
Volume	cubic inch (in.^{3}), cubic foot (ft^3), gallon (gal), barrel (bbl)
Speed/velocity	foot per second (ft/s)
Acceleration	foot per second per second (ft/s^2)
Density	slug per cubic foot (slugs/ft^3)
Specific weight	pound per cubic foot (lb/ft^3)
Specific volume	cubic foot per pound (ft^3/lb)
Dynamic viscosity	pound second per square foot (lb-s/ft^2)
Kinematic viscosity	square foot per second (ft^2/s)
Force	pound (lb)
Pressure	pound per square inch (lb/in.^{2} or psi)
Energy/work	foot pound (ft lb)
Quantity of heat	British thermal units (Btu)
Power	horsepower (HP)
Specific heat	Btu per pound per °F (Btu/lb/°F)
Thermal conductivity	Btu per hour per foot per °F (Btu/hr/ft/°F)

In SI units the following derived units are used:

Area	square meter (m^2)
Volume	cubic meter (m^3)
Speed/Velocity	meter/second (m/s)
Acceleration	meter per second per second (m/s^2)

Density	kilogram per cubic meter (kg/m^3)
Specific volume	cubic meter per kilogram (m^3/kg)
Dynamic viscosity	pascal second (Pa-s)
Kinematic viscosity	square meter per second (m^2/s)
Force	newton (N)
Pressure	newton per square meter or pascal (Pa)
Energy/work	newton meter or joule (J)
Quantity of heat	joule (J)
Power	joule per second or watt (W)
Specific heat	joule per kilogram per K (J/kg/K)
Thermal conductivity	joule/second/meter/kelvin (J/s/m/K)

Many other derived units are used in both English and SI units. A list of the more commonly used units in liquid pipeline hydraulics and their conversions is given in Appendix A.

2.2 Mass, Volume, Density, and Specific Weight

Several properties of liquids that affect liquid pipeline hydraulics will be discussed here. In steady-state hydraulics of liquid pipelines, the following properties are important: mass, volume, density, and specific weight.

2.2.1 Mass

Mass is defined as the quantity of matter. It is independent of temperature and pressure. Mass is measured in slugs (slugs) in English units or kilograms (kg) in SI units. In the past mass was used synonymously with weight. Strictly speaking weight depends upon the acceleration due to gravity at a certain geographic location and therefore is considered to be a force. Numerically mass and weight are interchangeable in the older FPS system of units. For example, a mass of 10 lbm is equivalent to a weight of 10 lbf. To avoid this confusion, in English units the slug has been adopted as the unit of mass. One slug is equal to 32.17 lb. Therefore, if a drum contains 55 gal of crude oil and weighs 410 lb, the mass of oil will be the same at any temperature and pressure. Hence the statement "conservation of mass."

2.2.2 Volume

Volume is defined as the space occupied by a given mass. In the case of the 55 gal drum above, 410 lb of crude oil occupies the volume of the drum. Therefore the crude oil volume is 55 gal. Consider a solid block of ice measuring 12 in. on each side. The volume of this block of ice is 12 × 12 × 12 or 1728 cubic inches or 1 cubic foot (ft^3). The volume of a certain petroleum product contained in a circular storage tank 100 ft in diameter and 50 ft high may be calculated as follows, assuming the liquid depth is 40 ft:

$$\text{Liquid volumes} = (\pi/4) \times 100 \times 100 \times 40 = 314{,}160 \text{ ft}^3$$

Liquids are practically incompressible, take the shape of their container and have a free surface. The volume of a liquid varies with temperature and pressure. However, since liquids are practically incompressible, pressure has negligible effect on volume. Thus, if the liquid volume measured at 50 pounds per square inch (psi) is 1000 gal, its volume at 1000 psi will not be appreciably different, provided the liquid temperature remained constant. Temperature, however, has a more significant effect on volume. For example, the 55 gal volume of liquid in a drum at a temperature of 60°F will increase to a slightly higher value (such as 56 gal) when the liquid temperature increases to 100°F. The increase in volume per unit temperature rise depends on the coefficient of expansion of the liquid. When measuring petroleum liquids, for the purpose of custody transfer, it is customary to correct volumes to a fixed temperature such as 60°F. Volume correction factors from American Petroleum Institute (API) publications are commonly used in the petroleum industry.

In the petroleum industry, it is customary to measure volume in gallons or barrels. One barrel is equal to 42 U.S. gallons. The Imperial gallon as used in the United Kingdom is a larger unit, approximately 20% larger than the U.S. gallon. In SI units volume is generally measured in cubic meters (m^3) or liters (L).

In a pipeline transporting crude oil or refined petroleum products, it is customary to talk about the "line fill volume" of the pipeline. The volume of liquid contained between two valves in a pipeline can be calculated simply by knowing the internal diameter of the pipe and the length of pipe between the two valves. By extension the total volume, or the line fill volume, of the pipeline can be easily calculated.

As an example, if a 16 in. pipeline, 0.250 in. wall thickness, is 5000 ft long from one valve to another, the line fill for this section of pipeline is

$$\begin{aligned}\text{Line fill volume} &= (\pi/4) \times (16 - 2 \times 0.250)^2 \times 5000 \\ &= 943{,}461.75 \text{ ft}^3 \text{ or } 168{,}038 \text{ bbl}\end{aligned}$$

The above calculation is based on conversion factors of:

1728 in.^{3} per ft^3
231 in.^{3} per gallon
42 gallons per barrel.

In Chapter 6 we discuss a simple formula for determining the line fill volume of a pipeline.

The volume flow rate in a pipeline is generally expressed in terms of cubic feet per second (ft^3/s), gallons per minute (gal/min), barrels per hour (bbl/hr), and barrels per day (bbl/day) in English units. In SI units, volume flow rate is measured in cubic meters per hour (m^3/hr) and liters per second (L/s).

It must be noted that since the volume of a liquid varies with temperature, the inlet flow rate and the outlet volume flow rate may be different in a long-distance pipeline, even if there are no intermediate injections or deliveries. This is due to the fact that the inlet flow rate may be measured at an inlet temperature of 70°F to be 5000 bbl/hr and the corresponding flow rate at the pipeline terminus, 100 miles away, may be measured at an outlet temperature different from the inlet temperature. The temperature difference is due to heat loss or gain between the pipeline liquid and the surrounding soil or ambient conditions. Generally, significant variation in temperature is observed when pumping crude oils or other products that are heated at the pipeline inlet. In refined petroleum products and other pipelines that are not heated, temperature variations along the pipeline are insignificant. In any case if the volume measured at the pipeline inlet is corrected to a standard temperature such as 60°F, the corresponding outlet volume can also be corrected to the same standard temperature. With temperature correction it can be assumed that the same flow rate exists throughout the pipeline from inlet to outlet, provided of course there are no intermediate injections or deliveries along the pipeline.

By the principle of conservation of mass, the mass flow rate at the inlet will equal that at the pipeline outlet since the mass of liquid does not change with temperature or pressure.

2.2.3 Density

Density of a liquid is defined as the mass per unit volume. Customary units for density are $slugs/ft^3$ in English units and kg/m^3 in SI units. This is also referred to as mass density. The weight density is defined as the weight per unit volume. This term is more commonly called specific weight and will be discussed next.

Since mass does not change with temperature or pressure, but volume varies with temperature, we can conclude that density will vary with temperature. Density and volume are inversely related since density is defined as mass per unit volume. Therefore, with increase in temperature liquid volume increases while its density decreases. Similarly, with reduction in temperature, liquid volume decreases and its density increases.

2.2.4 Specific Weight

Specific weight of a liquid is defined as the weight per unit volume. It is measured in lb/ft^3 in English units and N/m^3 in SI units.

If a 55 gal drum of crude oil weighs 410 lb (excluding the weight of the drum), the specific weight of crude oil is

$$(410/55) \text{ or } 7.45 \text{ lb/gal}$$

Similarly, consider the 5000 ft pipeline discussed in Section 2.2.2. The volume contained between the two valves was calculated to be 168,038 bbl. If we use the specific weight calculated above, we can estimate the weight of liquid contained in the pipeline as

$$7.45 \times 42 \times 168{,}038 = 52{,}579{,}090 \text{ lb or } 26{,}290 \text{ tons}$$

Similar to density, specific weight varies with temperature. Therefore, with increase in temperature specific weight will decrease. With reduction in temperature, liquid volume decreases and its specific weight increases.

The customary unit for specific weight is lb/ft^3 and lb/gal in the English units. The corresponding SI unit of specific weight is N/m^3.

For example, water has a specific weight of 62.4 lb/ft^3 or 8.34 lb/gal at 60°F. A typical gasoline has a specific weight of 46.2 lb/ft^3 or 6.17 lb/gal at 60°F.

Although density and specific weight are dimensionally different, it is common to use the term density instead of specific weight and vice versa when calculating hydraulics of liquid pipelines. Thus you will find that the density of water and specific weight of water are both expressed as 62.4 lb/ft^3.

2.3 Specific Gravity and API Gravity

Specific gravity of a liquid is the ratio of its density to the density of water at the same temperature and therefore has no units (dimensionless). It is a measure of how heavy a liquid is compared with water. By definition, the specific gravity of water is 1.00.

The term relative density is also used to compare the density of a liquid with another liquid such as water. It should be noted that both densities must be measured at the same temperature for the comparison to be meaningful.

At 60°F, a typical crude oil has a density of 7.45 lb/gal compared with a water density of 8.34 lb/gal. Therefore, the specific gravity of crude oil at 60°F is

$$\text{Specific gravity} = 7.45/8.34 \text{ or } 0.8933$$

Specific gravity, like density, varies with temperature. As temperature increases, both density and specific gravity decrease. Similarly, a decrease in temperature causes the density and specific gravity to increase in value. As with volume, pressure has very little effect on liquid specific gravity as long as pressures are within the range of most pipeline applications.

In the petroleum industry, it is customary to use units of °API for gravity. The API gravity is a laboratory-determined scale that compares the density of the liquid with that of water at 60°F (fixed at a value of 10). All liquids lighter than water will have API values higher than 10. Thus gasoline has an API gravity of 60 while a typical crude oil may be 35°API.

The API gravity versus the specific gravity relationship is as follows:

$$\text{Specific gravity (Sg)} = 141.5/(131.5 + \text{API}) \tag{2.1}$$

or

$$\text{API} = 141.5/\text{Sg} - 131.5 \tag{2.2}$$

Substituting an API value of 10 for water in Equation (2.1), yields as expected the specific gravity of 1.00 for water. It is seen from the above equation that the specific gravity of the liquid cannot be greater than 1.076 in order to result in a positive value of API.

Another scale of gravity for liquids heavier than water is known as the Baume scale. This scale is similar to the API scale with the exception of 140 and 130 being used in place of 141.5 and 131.5 respectively in Equations (2.1) and (2.2).

As another example, assume the specific gravity of gasoline at 60°F is 0.736. Therefore, the API gravity of gasoline can be calculated from Equation (2.2) as follows:

$$\text{API gravity} = 141.5/0.736 - 131.5 = 60.76°\text{API}$$

If diesel fuel is reported to have an API gravity of 35, the specific gravity can be calculated from Equation (2.1) as follows:

$$\text{Specific gravity} = 141.5/(131.5 + 35) = 0.8498$$

It must be noted that API gravity is always referred to at 60°F. Therefore in the Equations (2.1) and (2.2), specific gravity must also be measured at 60°F. Hence, it is meaningless to say that the API of a liquid is 35°API at 70°F.

API gravity is measured in the laboratory by the method described in ASTM D1298, using a properly calibrated glass hydrometer. The reader is referred to the *API Manual of Petroleum Measurements* for further discussion on API gravity.

2.3.1 Specific Gravity Variation with Temperature

It was mentioned earlier that the specific gravity of a liquid varies with temperature. It increases with a decrease in temperature and vice versa.

For the commonly encountered range of temperatures in liquid pipelines, the specific gravity of a liquid varies linearly with temperature. In other words, the specific gravity versus temperature can be expressed in the form of the following equation:

$$S_T = S_{60} - a(T - 60) \tag{2.3}$$

where

S_T = Specific gravity at temperature T
S_{60} = Specific gravity at 60°F
T = Temperature, °F
a = A constant that depends on the liquid

In Equation (2.3) the specific gravity S_T at temperature T is related to the specific gravity at 60°F by a straight-line relationship. Since the terms S_{60} and a are unknown quantities, two sets of specific gravities at two different temperatures are needed to determine the specific gravity versus temperature relationship. If the specific gravity at 60°F and the specific gravity at 70°F are known we can substitute these values in Equation (2.3) to obtain the unknown constant a. Once the value of a is known, we can easily calculate the specific gravity of the liquid at any other temperature using Equation (2.3). An example will illustrate how this is done.

Some handbooks such as the *Hydraulic Institute Engineering Data Book* and the *Flow of Fluids through Valves, Fittings and Pipes* (see References) provide specific gravity versus temperature curves from which the specific gravity of most liquids can be calculated at any temperature.

Example Problem 2.1

The specific gravity of gasoline at 60°F is 0.736. The specific gravity at 70°F is 0.729. What is the specific gravity at 50°F?

Solution

Using Equation (2.3), we can write

$$0.729 = 0.736 - a(70 - 60)$$

Solving for a, we get

$$a = 0.0007$$

We can now calculate the specific gravity at 50°F using Equation (2.3) as

$$S_{50} = 0.736 - 0.0007(50 - 60) = 0.743$$

2.3.2 Specific Gravity of Blended Liquids

Suppose a crude oil of specific gravity 0.895 at 70°F is blended with a lighter crude oil of specific gravity 0.815 at 70°F, in equal volumes. What will be the specific gravity of the blended mixture? Common sense suggests that since equal volumes are used, the resultant mixture should have a specific gravity that is the average of the two liquids, or

$$(0.895 + 0.815)/2 = 0.855$$

This is indeed the case, since specific gravity of a liquid is simply related to the mass and the volume of each liquid.

When two or more liquids are mixed homogeneously, the specific gravity of the resultant liquid can be calculated using the weighted average method. Thus, 10% of liquid A with a specific gravity of 0.85 when blended with 90% of liquid B that has a specific gravity of 0.89 results in a blended liquid with a specific gravity of

$$(0.1 \times 0.85) + (0.9 \times 0.89) = 0.886$$

It must be noted that when performing the above calculations, both specific gravities must be measured at the same temperature.

Using the above approach, the specific gravity of a mixture of two or more liquids can be calculated from the following equation:

$$S_b = \frac{(Q_1 \times S_1) + (Q_2 \times S_2) + (Q_3 \times S_3) + \cdots}{Q_1 + Q_2 + Q_3 + \cdots} \tag{2.4}$$

where

S_b = Specific gravity of the blended liquid
Q_1, Q_2, Q_3, etc. = Volume of each component
S_1, S_2, S_3, etc. = Specific gravity of each component

The above method of calculating the specific gravity of a mixture of two or more liquids cannot be directly applied when the gravities are expressed in °API values. If the component gravities of a mixture are given in °API we must first convert API values to specific gravities before applying Equation (2.4).

Example Problem 2.2

Three liquids A, B, and C are blended together in the ratio of 15%, 20%, and 65% respectively. Calculate the specific gravity of the blended liquid if the individual liquids have the following specific gravities at 70°F:

Specific gravity of liquid A: 0.815
Specific gravity of liquid B: 0.850
Specific gravity of liquid C: 0.895

Solution

Using Equation (2.4) we can calculate the specific gravity of the blended liquid as

$$S_b = (15 \times 0.815 + 20 \times 0.850 + 65 \times 0.895)/100 = 0.874$$

2.4 Viscosity

Viscosity is a measure of sliding friction between successive layers of a liquid that flows in a pipeline. Imagine several layers of liquid that constitute a flow between two fixed parallel horizontal plates. A thin layer adjacent to the bottom plate will be at rest or zero velocity. Each subsequent layer above this will have a different velocity compared with the layer below. This variation in the velocity of the liquid layers results in a velocity gradient. If the velocity is V at the layer that is located a distance of y from the bottom plate, the velocity gradient is approximately:

$$\text{Velocity gradient} = V/y \tag{2.5}$$

If the variation of velocity with distance is not linear, using calculus we can write more accurately that

$$\text{Velocity gradient} = \frac{dV}{dy} \tag{2.6}$$

where dV/dy represents the rate of change of velocity with distance or the velocity gradient.

Newton's law states that the shear stress between adjacent layers of a flowing liquid is proportional to the velocity gradient. The constant of proportionality is known as the absolute (or dynamic) viscosity of the liquid.

$$\text{Shear stress} = (\text{Viscosity})\,(\text{Velocity gradient})$$

The absolute viscosity of a liquid is measured in lb-s/ft^2 in English units and pascal-s in SI units. Other commonly used units of absolute viscosity are the poise and centipoise (cP).

The kinematic viscosity is defined as the absolute viscosity of a liquid divided by its density at the same temperature.

$$\nu = \mu/\rho \tag{2.7}$$

where

ν = Kinematic viscosity
μ = Absolute viscosity
ρ = Density

The units of kinematic viscosity are ft^2/s in English units and m^2/s in SI units. See Appendix A for conversion of units. Other commonly used units for kinematic viscosity are the stoke and centistoke (cSt). In the petroleum industry, two other units for kinematic viscosity are also used. These are saybolt seconds universal (SSU) and saybolt seconds furol (SSF). When expressed in these units, kinematic viscosity represents the time taken for a fixed volume of a liquid to flow through an orifice of defined size. Both absolute and kinematic viscosities vary with temperature. As temperature increases, liquid viscosity decreases and vice versa. However, unlike specific gravity, viscosity versus temperature is not a linear relationship. We will discuss this in Section 2.4.1.

Viscosity also varies somewhat with pressure. Significant variation in viscosity is found when pressures are several thousand psi. In most pipeline applications, viscosity of a liquid does not change appreciably with pressure.

For example, the viscosities of Alaskan North Slope (ANS) crude oil may be reported as 200 SSU at 60°F and 175 SSU at 70°F. Viscosity in SSU and SSF maybe converted to their equivalent in centistokes using the following equations:

Conversion from SSU to centistokes

$$\text{Centistokes} = 0.226(\text{SSU}) - 195/(\text{SSU}) \qquad \text{for } 32 \leq \text{SSU} \leq 100 \tag{2.8}$$

$$\text{Centistokes} = 0.220(\text{SSU}) - 135/(\text{SSU}) \qquad \text{for SSU} > 100 \tag{2.9}$$

Conversion from SSF to centistokes

$$\text{Centistokes} = 2.24(\text{SSF}) - 184/(\text{SSF}) \qquad \text{for } 25 < \text{SSF} \leq 40 \tag{2.10}$$

$$\text{Centistokes} = 2.16(\text{SSF}) - 60/(\text{SSF}) \qquad \text{for SSF} > 40 \tag{2.11}$$

Example Problem 2.3

Use the above equations to convert the viscosity of ANS crude oil from 200 SSU to its equivalent in centistokes.

Solution

Using Equation (2.9):

$$\begin{aligned} \text{Centistokes} &= 0.220 \times 200 - 135/200 \\ &= 43.33 \text{ cSt} \end{aligned}$$

The reverse process of converting from viscosity in cSt to its equivalent in SSU using Equations (2.8) and (2.9) is not quite so direct. Since Equations (2.8) and (2.9) are valid for certain ranges of SSU values, we need first to determine which of the two equations to use. This is difficult since the SSU value is unknown. Therefore we will have to assume that the SSU value to be calculated falls in one of the two ranges and proceed to calculate by trial and error. We will have to solve a quadratic equation to determine the SSU value for a given viscosity in cSt. An example will illustrate this method.

Example Problem 2.4

Suppose we are given a liquid viscosity of 15 cSt. Calculate the corresponding viscosity in SSU.

Solution

Let us assume that the calculated value in SSU is approximately $5 \times 15 =$ 75 SSU. This is a good approximation, since the SSU value is generally about 5 times the corresponding viscosity value in cSt. Since the assumed SSU value is 75, we need to use Equation (2.8) for converting between cSt and SSU.
Substituting 15 cSt in Equation (2.8) gives

$$15 = 0.226(\text{SSU}) - 195/(\text{SSU})$$

Replacing SSU with variable x, the above Equation becomes, after transposition

$$15x = 0.226x^2 - 195$$

Rearranging we get

$$0.226x^2 - 15x - 195 = 0$$

Solving for x, we get

$$x = [15 + (15 \times 15 + 4 \times 0.226 \times 195)^{1/2}]/(2 \times 0.226) = 77.5$$

Therefore, the viscosity is 77.5 SSU.

2.4.1 Viscosity Variation with Temperature

The viscosity of a liquid decreases as the liquid temperature increases and vice versa. For gases, the viscosity increases with temperature. Thus, if the viscosity of a liquid at 60°F is 35 cSt, as the temperature increases to 100°F, the viscosity could drop to a value of 15 cSt. The variation of liquid viscosity with temperature is not linear, unlike the variation of specific gravity with temperature discussed in Section 2.3.1. The variation of viscosity with temperature is found to be logarithmic in nature.
Mathematically, we can state the following:

$$\text{Log}_e(\nu) = A - B(T) \tag{2.12}$$

where

ν = Viscosity of liquid, cSt
T = Absolute temperature, °R or °K

$$T = (t + 460)°R \quad \text{if t is in °F} \tag{2.13}$$

$$T = (t + 273)°K \quad \text{if t is in °C} \tag{2.14}$$

A and B are constants that depend on the specific liquid.

It can be seen from Equation (2.12) that a graphic plot of $\log_e(\nu)$ against the temperature T will result in a straight line with a slope of $-B$. Therefore, if we have two sets of viscosity versus temperature values for a liquid we can determine the values of A and B by substituting the viscosity and temperature values in Equation (2.12). Once A and B are known we can calculate the viscosity at any other temperature using Equation (2.12). An example will illustrate this.

Example Problem 2.5

Suppose we are given the viscosities of a liquid at 60°F and 100°F as 43 cSt and 10 cSt. What is the viscosity at 80°F?

Solution

We will use Equation (2.12) to calculate the values of A and B first:

$$\text{Log}_e(43) = A - B(60 + 460)$$

and

$$\text{Log}_e(10) = A - B(100 + 460)$$

Solving the above two equations for A and B results in

$$A = 22.7232 \qquad B = 0.0365$$

Having found A and B, we can now calculate the viscosity of this liquid at any other temperature using Equation (2.12). To calculate the viscosity at 80°F:

$$\text{Log}_e(\nu) = 22.7232 - 0.0365(80 + 460) = 3.0132$$

Viscosity at 80°F = 20.35 cSt

In addition to Equation (2.12), several researchers have put forth various equations that attempt to correlate the viscosity variation of petroleum liquids with temperature. The most popular and accurate of the formulas is the one known as the ASTM method. In this method, also known as the ASTM D341 chart method, special graph paper with logarithmic scales is used to plot the viscosity of a liquid at two known temperatures. Once the two points are plotted on the chart and a line drawn connecting them, the viscosity at any intermediate temperature can be interpolated. To some extent, values of viscosity may also be extrapolated from this chart. This is shown in Figure 2.1.

In the following paragraphs we discuss how to calculate the viscosity variations with temperature, using the ASTM method, without using the special logarithmic graph paper.

$$\text{Log Log}(Z) = A - B\ \text{Log}(T) \tag{2.15}$$

where

Log = logarithm to base 10
Z depends on the viscosity of the liquid ν
ν = viscosity of liquid, cSt
T = Absolute temperature, °R or K

A and B are constants that depend on the specific liquid.
The variable Z is defined as follows:

$$Z = (\nu + 0.7 + C - D) \tag{2.16}$$

where C and D are

$$C = \exp[-1.14883 - 2.65868(\nu)] \tag{2.17}$$

$$D = \exp[-0.0038138 - 12.5645(\nu)] \tag{2.18}$$

C, D, and Z are all functions of the kinematic viscosity ν.

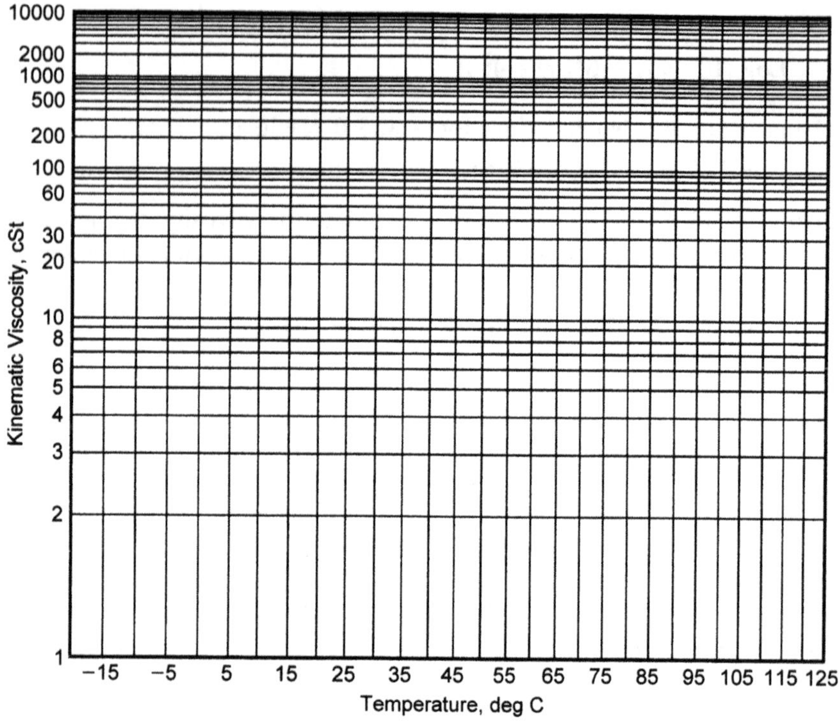

Figure 2.1 ASTM D341—viscosity temperature chart.

Given two sets of temperature viscosity values (T_1, ν_1) and (T_2, ν_2) we can calculate the corresponding values of C, D, and Z from Equations (2.17), (2.18), and (2.16).

We can then come up with two equations using the pairs of (T_1, Z_1) and (T_2, Z_2) values by substituting these values into Equation (2.15) as shown below:

$$\text{Log Log}(Z_1) = A - B\,\text{Log}(T_1) \tag{2.19}$$

$$\text{Log Log}(Z_2) = A - B\,\text{Log}(T_2) \tag{2.20}$$

From the above equations, the two unknown constants A and B can be easily calculated, since T_1, Z_1 and T_2, Z_2 values are known.

The following example will illustrate this approach for viscosity temperature variation.

Example Problem 2.6

A certain liquid has a temperature versus viscosity relationship as given below:

Temperature, °F	60	180
Viscosity, cSt	750	25

(a) Calculate the constants A and B that define the viscosity versus temperature correlation for this liquid using Equation (2.15).
(b) What is the estimated viscosity of this liquid at 85°F?

Solution

(a) At the first temperature 60°F, C, D, and Z are calculated using Equations (2.17), (2.18), and (2.16):

$$C_1 = \exp[-1.14883 - 2.65868 \times 750] = 0$$

$$D_1 = \exp[-0.0038138 - 12.5645 \times 750] = 0$$

$$Z_1 = (750 + 0.7) = 750.7$$

Similarly at the second temperature of 180°F, the corresponding values of C, D, and Z are calculated to be:

$$C_2 = \exp[-1.14883 - 2.65868 \times 25] = 0$$

$$D_2 = \exp[-0.0038138 - 12.5645 \times 25] = 0$$

$$Z_2 = (25 + 0.7) = 25.7$$

Substituting in Equation (2.19) we get

$$\text{Log Log}(750.7) = A - B\,\text{Log}(60 + 460)$$

or

$$0.4587 = A - 2.716B$$

$$\text{Log Log}(25.7) = A - B\,\text{Log}(180 + 460)$$

or

$$0.1492 = A - 2.8062B$$

Solving for A and B, we get

$$A = 9.778$$

$$B = 3.4313$$

(b) At a temperature of 85°F, using Equation (2.15) we get

$$\text{Log Log}(Z) = A - B\,\text{Log}(85 + 460)$$

$$\text{Log Log}(Z) = 9.778 - 3.4313 \times 2.7364 = 0.3886$$

$$Z = 279.78$$

Therefore,

$$\text{Viscosity at } 85^{\circ}\text{F} = 279.78 - 0.7 = 279.08 \text{ cSt}$$

2.4.2 Viscosity of Blended Products

Suppose a crude oil of viscosity 10 cSt at 60°F is blended with a lighter crude oil of viscosity 30 cSt at 60°F, in equal volumes. What will be the viscosity of the blended mixture? We cannot average the viscosities as we did with the specific gravities of mixtures earlier. This is due to the nonlinear nature of viscosity with mass and volumes of liquids.

When blending two or more liquids, the specific gravity of the blended product can be calculated directly, by using the weighted average approach as demonstrated in Section 2.3.2. However, the viscosity of a blend of two or more liquids cannot be calculated by simply using the ratio of each component. Thus if 20% of liquid A of viscosity 10 cSt is blended with 80% of liquid B with a viscosity of 30 cSt, the blended viscosity is *not* the following

$$0.2 \times 10 + 0.8 \times 30 = 26 \text{ cSt}$$

In fact, the actual blended viscosity would be 23.99 cSt as will be demonstrated in the following section.

The viscosity of a blend of two or more products can be estimated using the following equation:

$$\sqrt{V_b} = \frac{Q_1 + Q_2 + Q_3 + \cdots}{(Q_1/\sqrt{V_1}) + (Q_2/\sqrt{V_2}) + (Q_3/\sqrt{V_3})} \tag{2.21}$$

where

V_b = Viscosity of blend, SSU
Q_1, Q_2, Q_3, etc. = Volumes of each component
V_1, V_2, V_3, etc. = Viscosity of each component, SSU

Since Equation (2.19) requires the component viscosities to be in SSU, we cannot use this equation to calculate the blended viscosity when viscosity is less than 32 SSU (1.0 cSt).

Another method of calculating the viscosity of blended products has been in use in the pipeline industry for over four decades. This method is referred to as the Blending Index method. In this method a Blending Index is calculated for each liquid based on its viscosity. Next the Blending Index of the mixture is calculated from the individual blending indices by using the weighted average of the composition of the mixture. Finally, the viscosity of the blended mixture is calculated using the Blending Index of the mixture. The equations used are given below:

$$H = 40.073 - 46.414 \text{ Log}_{10} \text{ Log}_{10}(V + B) \tag{2.22}$$

$$B = 0.931(1.72)^V \qquad \text{for } 0.2 < V < 1.5 \tag{2.23}$$

$$B = 0.6 \qquad \text{for } V \geq 1.5 \tag{2.24}$$

$$Hm = [H1(pct1) + H2(pct2) + H3(pct3) + \cdots]/100 \tag{2.25}$$

where

H, H1, H2, ... = Blending Index of liquids
Hm = Blending Index of mixture
B = Constant in Blending Index equation
V = Viscosity, cSt
pct1, pct2, etc. = Percentage of liquids 1, 2, etc. in blended mixture.

Example Problem 2.7

Calculate the blended viscosity obtained by mixing 20% of liquid A with a viscosity of 10 cSt and 80% of liquid B with a viscosity of 30 cSt at 70°F.

Solution

First, convert the given viscosities to SSU to use Equation (2.21). The viscosity of liquid A is calculated using Equations (2.8) and (2.9):

$$10 = 0.226(V_A) - \frac{195}{V_A}$$

Rearranging we get

$$0.226\, V_A^2 - 10\, V_A - 195 = 0$$

Solving the quadratic equation for V_A we get

$$V_A = 58.90 \text{ SSU}$$

Similarly, viscosity of liquid B is

$$V_B = 140.72 \text{ SSU}$$

From Equation (2.21), the blended viscosity is

$$\sqrt{V_{\text{blnd}}} = \frac{20 + 80}{(20/\sqrt{58.9}) + (80/\sqrt{140.72})}$$
$$= 10.6953$$

Therefore the viscosity of the blend is

$$V_{\text{blnd}} = 114.39 \text{ SSU}$$

or

Viscosity of blend = 23.99 cSt after converting from SSU to cSt using Equation (2.9)

A graphical method is also available to calculate the blended viscosities of two petroleum products using ASTM D341. This method involves using a logarithmic chart with viscosity scales on the left and right side of the paper. The horizontal axis is for selecting the percentage of each product as shown in Figure 2.2. This chart is also available in handbooks such as *Flow of Fluids through Valves, Fittings and Pipes* and the *Hydraulic Institute Engineering Data Book* (see References). It must be noted that the viscosities of both products must be plotted at the same temperature.

Using this method, the blended viscosity of two products at a time is calculated and the process repeated for multiple products. Thus if three products are blended in the ratios of 10%, 20%, and 70%, we would first calculate the viscosity of the blend using the first two liquids, considering 10 parts of liquid A mixed with 20 parts of liquid B. This means that the blend would be calculated on the basis of one-third of liquid A and two-thirds of liquid B. Next this blended liquid will be mixed with liquid C in the proportions of 30% and 70% respectively.

2.5 Vapor Pressure

Vapor pressure of a liquid is defined as the pressure at a given temperature at which the liquid and vapor exist in equilibrium. The normal boiling point of a liquid can thus be defined as the temperature at which the vapor pressure equals the atmospheric pressure. In the laboratory the vapor pressure is measured at a fixed temperature of 100°F and is then reported as the Reid vapor pressure. The vapor pressure of a liquid increases with temperature. Charts are available to determine the actual vapor pressure of a liquid at any temperature once its Reid vapor pressure is known. The

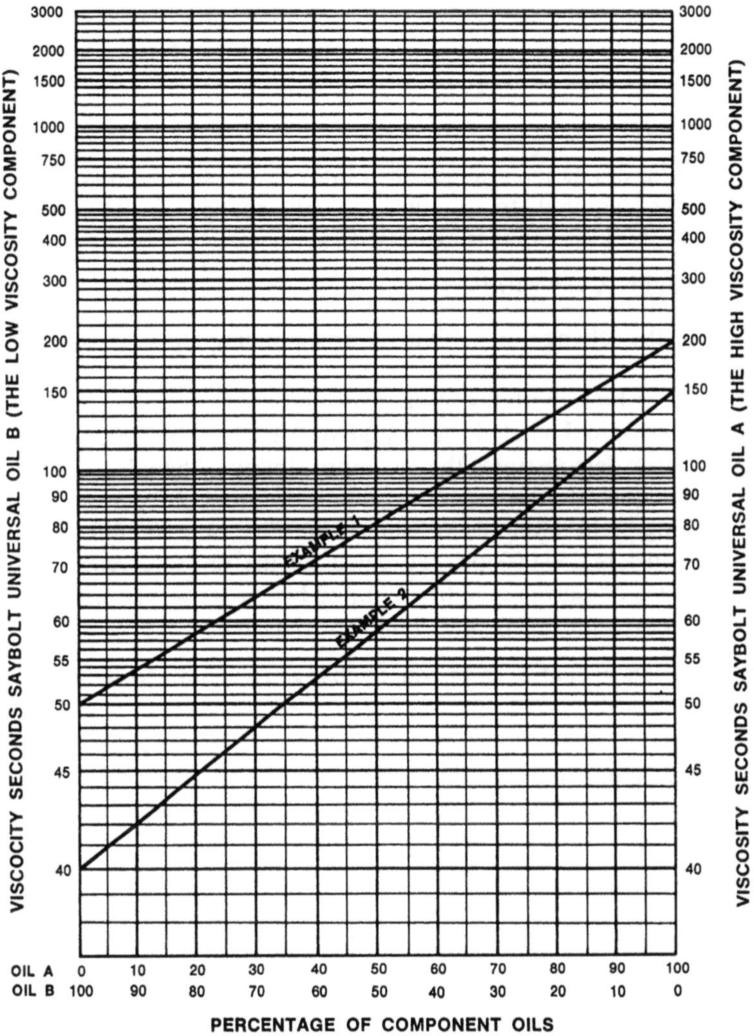

Figure 2.2 Viscosity blending chart.

reader is referred to *Fluid Flow through Valves, Fittings and Pipes* for vapor pressure charts.

The importance of vapor pressure will be evident when we discuss the operation of centrifugal pumps on pipelines. To prevent cavitation of pumps, the liquid vapor pressure at the flowing temperature must be taken into account in the calculation of net positive suction head (NPSH) available at the pump suction. Centrifugal pumps are discussed in Chapter 7.

2.6 Bulk Modulus

The bulk modulus of a liquid is a measure of its compressibility. It is defined as the pressure required to produce a unit change in its volume. Mathematically, bulk modulus is expressed as

$$\text{Bulk modulus } K = VdP/dV \tag{2.26}$$

where dV is the change in volume corresponding to a change in pressure of dP.

The units of bulk modulus, K, are psi or kPa. For most liquids the bulk modulus is approximately in the range of 250,000 to 300,000 psi. The fairly high number demonstrates the incompressibility of liquids.

Let us demonstrate the incompressibility of liquids by performing a calculation using bulk modulus. Assume that the bulk modulus of a petroleum product is 250,000 psi. To calculate the pressure required to change the volume of a given quantity of liquid by 1% we would proceed as follows. From Equation (2.26), with some rearrangement,

$$\text{Bulk modulus} = \text{change in pressure}/(\text{change in volume}/\text{volume})$$

Therefore

$$250{,}000 = \text{change in pressure}/(0.01)$$

Therefore

$$\text{Change in pressure} = 2500 \text{ psi}$$

It can be seen from the above that a fairly large pressure is required to produce a very small (1%) change in the liquid volume. Hence we say that liquids are fairly incompressible.

Bulk modulus is used in line pack calculations and transient flow analysis. There are two bulk modulus values used in practice: isothermal and adiabatic. The bulk modulus of a liquid depends on temperature, pressure, and specific gravity. The following empirical equations, also known as ARCO formulas, may be used to calculate the bulk modulus.

2.6.1 Adiabatic Bulk Modulus

$$Ka = A + B(P) - C(T)^{1/2} - D(API) - E(API)^2 + F(T)(API) \tag{2.27}$$

where

$A = 1.286 \times 10^6$
$B = 13.55$
$C = 4.122 \times 10^4$
$D = 4.53 \times 10^3$
$E = 10.59$
$F = 3.228$
P = Pressure, psig
T = Temperature, °R
API = API gravity of liquid

2.6.2 Isothermal Bulk Modulus

$$Ki = A + B(P) - C(T)^{1/2} + D(T)^{3/2} - E(API)^{3/2} \tag{2.28}$$

where

$A = 2.619 \times 10^6$
$B = 9.203$
$C = 1.417 \times 10^5$
$D = 73.05$
$E = 341.0$
P = Pressure, psig
T = Temperature, °R
API = API gravity of liquid

For a typical crude oil of 35°API gravity at 1000 psig pressure and 80°F temperature, the bulk modulus values calculated from Equations (2.27) and (2.28) are:

Adiabatic bulk modulus = 231,426 psi
Isothermal bulk modulus = 181,616 psi

The bulk modulus of water at 70°F is 320,000 psi.

2.7 Fundamental Concepts of Fluid Flow

In this section we discuss fundamental concepts of fluid flow that will set the stage for the succeeding chapter. The basic principles of continuity and energy equations are introduced first.

2.7.1 Continuity

One of the fundamental concepts that must be satisfied in any type of pipe flow is the principle of continuity of flow. This principle states that the total amount of fluid passing through any section of a pipe is fixed. This may also be thought of as the principle of conservation of mass. Basically, it means that liquid is neither created nor destroyed as it flows through a pipeline. Since mass is the product of the volume and density, we can write the following equation for continuity:

$$M = \text{Vol} \times \rho = \text{Constant} \tag{2.29}$$

where

M = Mass flow rate at any point in the pipeline, slugs/s
Vol = Volume flow rate at any point in the pipeline, ft^3/s
ρ = Density of liquid at any point in the pipeline, $slugs/ft^3$

Since the volume flow rate at any point in a pipeline is the product of the area of cross-section of the pipe and the average liquid velocity, we can rewrite Equation (2.29) as follows:

$$M = A \times V \times \rho = \text{Constant} \tag{2.30}$$

where

M = Mass flow rate at any point in the pipeline, slugs/s
A = Area of cross-section of pipe, ft^2
V = Average liquid velocity, ft/s
ρ = Density of liquid at any point in the pipeline, $slugs/ft^3$

Since liquids are generally considered to be incompressible and therefore density does not change appreciably, the continuity equation reduces to

$$AV = \text{Constant} \tag{2.31}$$

2.7.2 Energy Equation

The basic principle of conservation of energy applied to liquid hydraulics is embodied in Bernoulli's equation, which simply states that the total energy of the fluid contained in the pipeline at any point is a constant. Obviously, this is an extension of the principle of conservation of energy which states that energy is neither created nor destroyed, but transformed from one form to another.

Consider the pipeline shown in Figure 2.3 that depicts flow from point A to point B with the elevation of point A being Z_A and elevation at B being Z_B above some chosen datum. The pressure in the liquid at point A is P_A and that at B is P_B. Assuming a general case, where the pipe diameter at A may be different from that at B, we will designate the velocities at A and B to be V_A and V_B respectively. Consider a particle of the liquid of

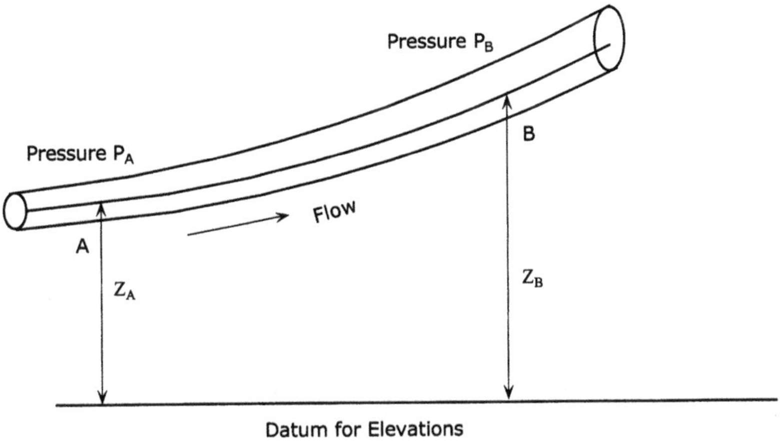

Figure 2.3 Energy of a liquid in pipe flow.

weight W at point A in the pipeline. This liquid particle at A may be considered to possess a total energy E that consists of three components:

Energy due to position, or potential energy $= WZ_A$

Energy due to pressure, or pressure energy $= WP_A/\gamma$

Energy due to velocity, or kinetic energy $= W(V_A^2)/2g$

where

γ = Specific weight of the liquid

We can thus state that

$$E = WZ_A + WP_A/\gamma + WV_A^2/2g \quad (2.32)$$

Dividing by W throughout, we get the total energy per unit weight of liquid as

$$H_A = Z_A + P_A/\gamma + V_A^2/2g \quad (2.33)$$

where

H_A = total energy per unit weight at point A

Considering the same liquid particle as it arrives at point B, the total energy per unit weight at B is

$$H_B = Z_B + P_B/\gamma + V_B^2/2g \quad (2.34)$$

Due to conservation of energy

$$H_A = H_B$$

Therefore,

$$Z_A + P_A/\gamma + V_A^2/2g = Z_B + P_B/\gamma + V_B^2/2g \tag{2.35}$$

Equation (2.35) is one form of Bernoulli's equation for fluid flow.

In real-world pipeline transportation, there is energy loss between point A and point B, due to friction in the pipe. We include the energy loss due to friction by modifying Equation (2.35) as follows:

$$Z_A + P_A/\gamma + V_A^2/2g = Z_B + P_B/\gamma + V_B^2/2g + \Sigma h_L \tag{2.36}$$

where

Σh_L = all the head losses between points A and B, due to friction

In Bernoulli's equation (2.35), we must also include any energy added to the liquid, such as when there is a pump between points A and B. Thus the left-hand side of the equation will have a positive term added to it that will represent the energy generated by a pump. Equation (2.36) will be modified as follows to include a pump at point A that will add a certain amount of pump head to the liquid:

$$Z_A + P_A/\gamma + V_A^2/2g + H_P = Z_B + P_B/\gamma + V_B^2/2g + \Sigma h_L \tag{2.37}$$

where

H_P = pump head added to the liquid at point A

In Chapter 3, we further explore the concepts of pressure, velocity, flow rates, and energy lost due to pipe friction.

2.8 Summary

In this chapter we discussed the more important properties of liquids that determine the nature of liquid flow in pipelines. The specific gravity and viscosity of liquids were explained along with how to calculate these properties in liquid mixtures and at various temperatures. We also introduced the basic concepts of liquid flow consisting of the continuity equation and the energy equation embodied in Bernoulli's equation.

2.9 Problems

2.9.1 Calculate the specific weight and specific gravity of a liquid that weighs 312 lb, contained in a volume of 5.9 ft^3. Assume water weighs 62.4 lb/ft^3.

2.9.2 The specific gravities of a liquid at 60°F and 100°F are reported to be 0.895 and 0.815 respectively. Determine the specific gravity of the liquid at 85°F. Assume a linear relationship between specific gravity and temperature.

2.9.3 The gravity of a petroleum product is 59°API. Calculate the corresponding specific gravity at 60°F.

2.9.4 The viscosity of a liquid at 70°F is 45 cSt. Express this viscosity in SSU. If the specific gravity at 70°F is 0.885, determine the absolute or dynamic viscosity.

2.9.5 The viscosities of a crude oil at 60°F and 100°F are 40 cSt and 15 cSt respectively. Using the ASTM correlation method, calculate the viscosity of this product at 80°F.

2.9.6 Two liquids are blended to form a homogeneous mixture. Liquid A has a specific gravity of 0.815 at 70°F and a viscosity of 15 cSt at 70°F. At the same temperature, liquid B has a specific gravity of 0.85 and viscosity of 25 cSt. If 20% of liquid A is blended with 80% of liquid B, calculated the specific gravity and viscosity of the blended product.

2.9.7 Using the viscosity blending chart, calculate the blended viscosity for two liquids as follows:

Product	Percentage	Viscosity (SSU)
Liquid A	15	50
Liquid B	85	200

2.9.8 If liquid A with a viscosity of 40 SSU is blended with liquid B of viscosity 150 SSU, what percentage of each component would be required to obtain a blended viscosity of 46 SSU?

2.9.9 In Figure 2.3, consider the pipe to be 20 in. diameter. The liquid is water with a specific gravity of 1.00. The point A is at elevation 100 ft and B is at elevation 200 ft. The pressure at A is 500 psi and that at B is 400 psi. Specific weight of water is 62.34 lb/ft^3. Write down the Bernoulli equation for energy conservation between points A and B.

3

Pressure Drop due to Friction

In this chapter, we introduce the concept of pressure in a liquid and how it is measured. The liquid flow velocity in a pipe, types of flow, and the importance of the Reynolds number will be discussed. For different flow regimes, such as laminar, critical, or turbulent, methods will be discussed as to how to calculate the pressure drop due to friction. Several popular formulas such as the Colebrook-White and Hazen-Williams equations will be presented and compared. Also we will cover minor pressure losses in piping such as those due to fittings and valves, and those resulting from pipe enlargements and contractions. We will also explore drag reduction as a means of reducing energy loss in pipe flow.

3.1 Pressure

Hydrostatics is the study of hydraulics that deals with liquid pressures and forces resulting from the weight of the liquid at rest. Although this book is mainly concerned with flow of liquids in pipelines, we will address some issues related to liquids at rest in order to discuss some fundamental issues pertaining to liquids at rest and in motion.

The force per unit area at a certain point within a liquid is called the pressure, p. This pressure at a certain depth, h, below the free surface of the liquid consists of equal pressures in all directions. This is known as Pascal's law. Consider an imaginary flat surface within the liquid located at a depth, h, below the liquid surface as shown in Figure 3.1. The pressure on this surface must act normal to the surface at all points along the surface because liquids at rest cannot transmit shear. The variation of pressure with the depth of the liquid is calculated by considering forces acting on a thin vertical cylinder of height Δh and a cross-sectional area Δa as shown in Figure 3.1.

Since the liquid is at rest, the cylindrical volume is in equilibrium due to the forces acting upon it. By the principles of statics, the algebraic sum of all forces acting on this cylinder in the vertical and horizontal directions must equal zero. The vertical forces on the cylinder consists of the weight of the cylinder and the forces due to liquid pressure P_1 at the top and P_2 at the bottom, as shown in Figure 3.1. Since the specific weight of the liquid, γ, does not change with pressure, we can write the following equation for the summation of forces in the vertical direction:

$$P_2 \Delta a = \gamma \Delta h \Delta a + P_1 \Delta a$$

where the term $\gamma \Delta h \Delta a$ represents the weight of the cylindrical element. Simplifying the above we get

$$P_2 = \gamma \Delta h + P_1 \tag{3.1}$$

If we now imagine that the cylinder is extended to the liquid surface, P_1 becomes the pressure at the liquid surface (atmospheric pressure P_a) and Δh becomes h, the depth of the point in the liquid where the pressure is P_2. Replacing P_2 with P, the pressure in the liquid at depth h, Equation (3.1)

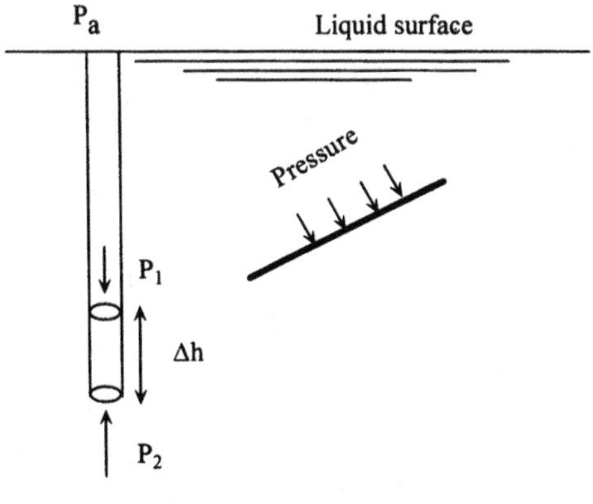

Figure 3.1 Pressure in a liquid.

becomes

$$P = \gamma h + P_a \tag{3.2}$$

From Equation (3.2) we conclude that the pressure in a liquid at a depth h increases with the depth. If the term P_a (atmospheric pressure) is neglected we can state that the gauge pressure (based on zero atmospheric pressure) at a depth h is simply γh. Therefore, the gauge pressure is

$$P = \gamma h \tag{3.3}$$

Dividing both sides by γ and transposing we can write

$$h = P/\gamma \tag{3.4}$$

In Equation (3.4) the term h represents the "pressure head" corresponding to the pressure P. It represents the depth in feet of liquid of specific weight γ required to produce the pressure P. Values of absolute pressure $(P + P_a)$ are always positive whereas the gauge pressure P may be positive or negative depending on whether the pressure is greater or less than the atmospheric pressure. Negative gauge pressure means that a partial vacuum exists in the liquid.

From the above discussion it is clear that the absolute pressure within a liquid consists of the head pressure due to the depth of liquid and the atmospheric pressure at the liquid surface. The atmospheric pressure at a geographic location varies with the elevation above sea level. Because the density of the atmospheric air varies with the altitude, a straight-line relationship does not exist between the altitude and the atmospheric pressure (unlike the linear relationship between liquid pressure and depth). For most purposes, we can assume that the atmospheric pressure at sea level is approximately 14.7 psi in English units, or approximately 101 kPa in SI units.

The instrument used to measure the atmospheric pressure at a given location is called a barometer. A typical barometer is shown in Figure 3.2. In such an instrument the tube is filled with a heavy liquid (usually mercury) then quickly inverted and positioned in a container full of the liquid as shown in Figure 3.2. If the tube is sufficiently long, the level of liquid will fall slightly to cause a vapor space at the top of the tube just above the liquid surface. Equilibrium will be reached when the liquid vaporizes in the vapor space and creates a pressure P_v. Because the density of mercury is high (approximately 13 times that of water) and its vapor pressure is low, it is an ideal liquid for a barometer. If a liquid such as water were used, a rather long tube would be needed to measure the atmospheric pressure, as we shall see shortly.

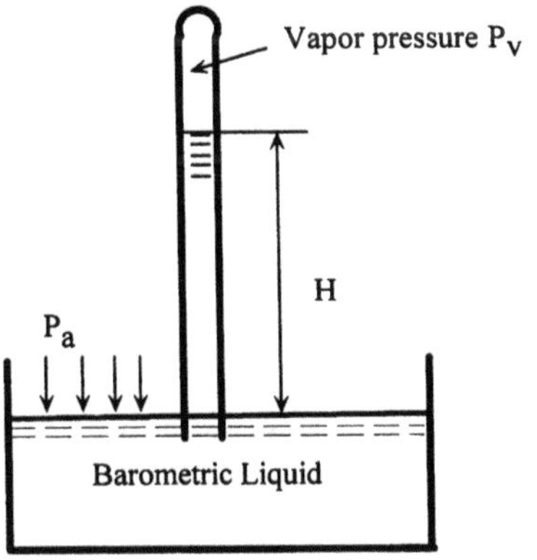

Figure 3.2 Barometer for measuring pressure.

From Figure 3.2 the atmospheric pressure P_a exerted at the surface of the liquid is equal to the sum of the vapor pressure P_v and the pressure generated by the column of the barometric liquid of height H_b.

$$P_a = P_v + \gamma H_b \tag{3.5}$$

where

P_a = Atmospheric pressure
P_v = Vapor pressure of barometric liquid
γ = Specific weight of barometric liquid
H_b = Barometric reading

In Equation (3.5), if pressures are in psi and liquid specific weight is in lb/ft^3, the pressures must be multiplied by 144 to obtain the barometric reading in feet of liquid.

Equation (3.5) is valid for barometers with any liquid. Since the vapor pressure of mercury is negligible we can rewrite Equation (3.5) for a mercury barometer as follows:

$$P_a = \gamma H_b \tag{3.6}$$

Let us compare the use of water and mercury as barometric liquids to measure the atmospheric pressure.

Example Problem 3.1

Assume the vapor pressure of water at 70°F is 0.3632 psi and its specific weight is 62.3 lb/ft^3. Mercury has a specific gravity of 13.54 and negligible vapor pressure. The sea level atmospheric pressure is 14.7 psi. Determine the barometric heights for water and mercury.

Solution

From Equation (3.5) for water,

$$14.7 = 0.3632 + (62.3/144)\ H_b$$

The barometric height for water is

$$H_b = (14.7 - 0.3632) \times 144/62.3 = 33.14\ \text{ft}$$

Similarly, for mercury, neglecting the vapor pressure, using Equation (3.6) we get

$$14.7 \times 144 = (13.54 \times 62.3)H_b$$

The barometric height for mercury is

$$H_b = (14.7 \times 144)/(13.54 \times 62.3) = 2.51\ \text{ft}$$

It can be seen from the above that the mercury barometer requires a much shorter tube than a water barometer.

Manometers are instruments used to measure pressure in reservoirs, channels, and pipes. Manometers are discussed further in Chapter 10.

The pressure in a liquid is measured in lb/in.^{2} (psi) in English units or kilopascal (kPa) in SI units. Since pressure is measured using a gauge and is relative to the atmospheric pressure at the specific location, it is also reported as psig (psi gauge). The absolute pressure in a liquid is the sum of the gauge pressure and the atmospheric pressure at the location. Thus,

$$\text{Absolute pressure in psia} = \text{Gauge pressure in psig} + \text{Atmospheric pressure}$$

For example, if the pressure gauge reading is 800 psig, the absolute pressure in the liquid is

$$P_{abs} = 800 + 14.7 = 814.7\ \text{psia}$$

This is based on the assumption that atmospheric pressure at the location is 14.7 psia.

Pressure in a liquid may also be referred to in terms of feet (or meters in SI units) of liquid head. By dividing the pressure in lb/ft^2 by the liquid specific weight in lb/ft^3, we get the pressure head in feet of liquid. When expressed this way, the head represents the height of the liquid column required to match the given pressure in psig. For example, if the pressure in a liquid is 1000 psig, the head of liquid corresponding to this pressure is calculated as follows:

$$\text{Head} = 2.31(\text{psig})/\text{Spgr} \qquad \text{ft (English units)} \tag{3.7}$$

$$\text{Head} = 0.102(\text{kPa})/\text{Spgr} \qquad \text{m (SI units)} \tag{3.8}$$

where

Spgr = Liquid specific gravity

The factor 2.31 in Equation (3.7) comes from the ratio

$$\frac{144\ \text{in.}^2/\text{ft}^2}{62.34\ \text{lb/ft}^3}$$

where 62.34 lb/ft^3 is the specific weight of water. Therefore, if the liquid specific gravity is 0.85, the equivalent liquid head is

$$\text{Head} = (1000)(2.31)/0.85 = 2717.65\ \text{ft}$$

This means that the liquid pressure of 1000 psig is equivalent to the pressure exerted at the bottom of a liquid column, of specific gravity 0.85, that is 2717.65 feet high. If such a column of liquid had a cross-sectional area of 1 square inch, the weight of the column would be

$$2717.65(1/144)(62.34)(0.85) = 1000\ \text{lb}$$

where 62.34 lb/ft^3 is the density of water. As the above weight acts on an area of 1 square inch, the calculated pressure is therefore 1000 psig.

We can analyze head pressure due to a column of liquid in another way: Consider a cylindrical column of liquid, of height H ft and cross-sectional area A in.^2. If the top surface of the liquid column is open to the atmosphere, we can calculate the pressure exerted by this column of liquid at its base as

$$\text{Pressure} = \frac{\text{Weight of liquid column}}{\text{Area of cross-section}}$$

or

$$\text{Pressure} = \frac{(\text{Volume} \times \text{Specific weight})}{\text{Area of cross-section}}$$

$$= (AH\gamma)/(144 \times A)$$

or

$$\text{Pressure} = H\gamma/144 \tag{3.9}$$

where

γ = Specific weight of liquid, lb/ft^3

The factor 144 is used to convert from in.^2 to ft^2.

If we use 62.34 lb/ft^3 for the specific weight of water, the pressure of a water column calculated from Equation (3.9) is

$$\text{Pressure} = H \times 62.34/144 = H/2.31$$

3.2 Velocity

Velocity of flow in a pipeline is the average velocity based on the pipe diameter and liquid flow rate. It may be calculated as follows:

$$\text{Velocity} = \text{Flow rate}/\text{Area of flow}$$

Depending on the type of flow (laminar, turbulent, etc.), the liquid velocity in a pipeline at a particular pipe cross-section will vary along the pipe radius. The liquid molecules at the pipe wall are at rest and therefore have zero velocity. As we approach the centerline of the pipe, the liquid molecules are increasingly free and therefore have increasing velocity. The variations in velocity for laminar flow and turbulent flow are as shown in Figure 3.3. In laminar flow (also known as viscous or streamline flow), the variation in velocity at a pipe cross-section is parabolic. In turbulent flow there is an approximate trapezoidal shape to the velocity profile.

If the units of flow rate are bbl/day and pipe inside diameter is in inches the following equation for average velocity may be used:

$$V = 0.0119(\text{bbl/day})/D^2 \tag{3.10}$$

where

V = Velocity, ft/s

D = Pipe internal diameter, in.

Other forms of the equation for velocity in different units are as follows:

$$V = 0.4085(\text{gal/min})/D^2 \tag{3.11}$$

$$V = 0.2859(\text{bbl/hr})/D^2 \tag{3.12}$$

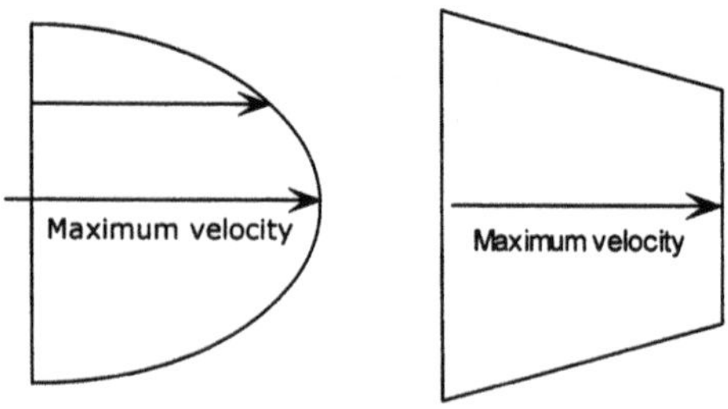

Figure 3.3 Velocity variation in a pipe for laminar flow (left) and turbulent flow (right).

where

V = Velocity, ft/s

D = Pipe internal diameter, in.

In SI units, the velocity is calculated as follows:

$$V = 353.6777(m^3/hr)/D^2 \tag{3.13}$$

where

V = Velocity, m/s

D = Pipe internal diameter, mm

For example, liquid flowing through a 16 in. pipeline (wall thickness 0.250 in.) at the rate of 100,000 bbl/day has an average velocity of

$$0.0119(100{,}000)/(15.5)^2 = 4.95 \text{ ft/s}$$

This represents the average velocity at a particular cross-section of pipe. The velocity at the centerline will be higher than this, depending on whether the flow is turbulent or laminar.

3.3 Reynolds Number

Flow in a liquid pipeline may be smooth, laminar flow (also known as viscous or streamline flow). In this type of flow the liquid flows in layers or laminations without causing eddies or turbulence. If the pipe were transparent and we injected a dye into the flowing stream, it would flow smoothly in a straight line confirming smooth or laminar flow. As the liquid flow rate is increased, the velocity increases and the flow will change from laminar flow to turbulent flow with eddies and disturbances. This can be seen clearly when a dye is injected into the flowing stream.

An important dimensionless parameter called the Reynolds number is used in classifying the type of flow in pipelines. The Reynolds number of flow, R, is calculated as follows:

$$R = VD\rho/\mu \tag{3.14}$$

where

V = Average velocity, ft/s

D = Pipe internal diameter, ft

ρ = Liquid density, slugs/ft^3

μ = Absolute viscosity, lb-s/ft^2

R = Reynolds number, dimensionless

Since the kinematic viscosity $\nu = \mu/\rho$ the Reynolds number can also be expressed as

$$R = VD/\nu \tag{3.15}$$

where

ν = Kinematic viscosity, ft^2/s

Care should be taken to ensure that proper units are used in Equations (3.14) and (3.15) such that R is dimensionless.

Flow through pipes is classified into three main flow regimes:

1. Laminar flow: $R < 2000$
2. Critical flow: $R > 2000$ and $R < 4000$
3. Turbulent flow: $R > 4000$

Depending upon the Reynolds number, flow through pipes will fall in one of the above three flow regimes. Let us first examine the concepts of Reynolds number. Sometimes an R value of 2100 is used as the limit of laminar flow.

Using the customary units of the pipeline industry, the Reynolds number can be calculated using the following formula:

$$R = 92.24\ Q/(\nu D) \tag{3.16}$$

where

Q = Flow rate, bbl/day

D = Internal diameter, in.

ν = Kinematic viscosity, cSt

Equation (3.16) is simply a modified form of Equation (3.15) after performing conversions to commonly used pipeline units. R is still a dimensionless value.

Another version of the Reynolds number in English units is as follows:

$$R = 3160\ Q/(\nu D) \tag{3.17}$$

where

Q = Flow rate, gal/min

D = Internal diameter, in.

ν = Kinematic viscosity, cSt

An equivalent equation for Reynolds number in SI units is

$$R = 353{,}678\ Q/(\nu D) \tag{3.18}$$

where

Q = Flow rate, m^3/h

D = Internal diameter, mm

ν = Kinematic viscosity, cSt

As indicated earlier, if the Reynolds number is less than 2000, the flow is considered laminar. This means that the various layers of liquid flow without turbulence in the form of laminations. We now illustrate the various flow regimes using an example.

Consider a 16 in. pipeline, with a wall thickness of 0.250 in., transporting a liquid of viscosity 250 cSt. At a flow rate of 50,000 bbl/day the Reynolds number is, using Equation (3.16),

$$R = 92.24(50{,}000)/(250 \times 15.5) = 1190$$

Since R is less than 2000, this flow is laminar. If the flow rate is tripled to 150,000 bbl/day, the Reynolds number becomes 3570 and the flow will be in the critical region. At flow rates above 168,040 bbl/day the Reynolds number exceeds 4000 and the flow will be in the turbulent region. Thus, for this 16 in. pipeline and given liquid viscosity of 250 cSt, flow will be fully turbulent at flow rates above 168,040 bbl/day.

As the flow rate and velocity increase, the flow regime changes. With changes in flow regime, the energy lost due to pipe friction increases. At laminar flow, there is less frictional energy lost compared with turbulent flow.

3.4 Flow Regimes

In summary, the three flow regimes may be distinguished as follows:

Laminar: Reynolds number < 2000
Critical: Reynolds number > 2000 and Reynolds number < 4000
Turbulent: Reynolds number > 4000

As liquid flows through a pipeline, energy is lost due to friction between the pipe surface and the liquid and due to the interaction between liquid molecules. This energy lost is at the expense of liquid pressure. (See Equation (2.37), Bernoulli's equation, in Chapter 2.) Hence we refer to the frictional energy lost as the pressure drop due to friction.

The pressure drop due to friction in a pipeline depends on the flow rate, pipe diameter, pipe roughness, liquid specific gravity, and viscosity. In addition, the frictional pressure drop depends on the Reynolds number (and hence the flow regime). Our objective would be to calculate the pressure drop given these pipe and liquid properties and the flow regime.

The pressure drop due to friction in a given length of pipe, expressed in feet of liquid head (h), can be calculated using the Darcy-Weisbach equation as follows:

$$h = f(L/D)(V^2/2g) \tag{3.19}$$

where

f = Darcy friction factor, dimensionless, usually a number between 0.008 and 0.10
L = Pipe length, ft
D = Pipe internal diameter, ft

V = Average liquid velocity, ft/s
g = Acceleration due to gravity, 32.2 ft/s^2 in English units

In laminar flow, the friction factor f depends only on the Reynolds number. In turbulent flow f depends on pipe diameter, internal pipe roughness, and Reynolds number, as we will see shortly.

Example Problem 3.2

Consider a pipeline transporting 4000 bbl/hr of gasoline (Spgr = 0.736). Calculate the pressure drop in a 5000 ft length of 16 in. pipe (wall thickness 0.250 in.) using the Darcy-Weisbach equation. Assume the friction factor is 0.02.

Solution

Using Equation (3.10):

$$\text{Average liquid velocity} = 0.0119(4000 \times 24)/(15.5)^2 = 4.76 \text{ ft/s}$$

Using the Darcy-Weisbach equation (3.19):

$$\text{Pressure drop} = 0.02(5000)(12/15.5)(4.76^2/64.4) = 27.24 \text{ ft of head}$$

Converting to pressure in psi, using Equation (3.7):

$$\text{Pressure drop} = 27.24(0.736)/2.31 = 8.68 \text{ psi}$$

In the above calculations, the friction factor f was assumed to be 0.02. However, the actual friction factor for a particular flow depends on various factors as explained previously. In the next section, we will see how the friction factor is calculated for the various flow regimes.

3.5 Friction Factor

For laminar flow, with Reynolds number $R < 2000$, the Darcy friction factor f is calculated from the simple relationship

$$f = 64/R \tag{3.20}$$

It can be seen from Equation (3.20) that for laminar flow the friction factor depends only on the Reynolds number and is independent of the internal condition of the pipe. Thus, regardless of whether the pipe is smooth or rough, the friction factor for laminar flow is a number that varies inversely with the Reynolds number. Therefore, if the Reynolds number $R = 1800$, the friction factor becomes

$$f = 64/1800 = 0.0356$$

It might appear that, since f for laminar flow decreases with Reynolds number, then from the Darcy-Weisbach equation the pressure drop will decrease with an increase in flow rate. This is not true. Since pressure drop is proportional to the square of the velocity V (Equation 3.19), the influence of V is greater than that of f. Therefore, pressure drop will increase with flow rate in the laminar region.

To illustrate, consider the Reynolds number example in Section 3.3 above. If the flow rate is increased from 50,000 bbl/day to 80,000 bbl/day, the Reynolds number R will increase from 1190 to 1904 (still laminar). The velocity will increase from V_1 to V_2 as follows:

$$V_1 = 0.0119(50{,}000)/(15.5)^2 = 2.48 \text{ ft/s}$$

$$V_2 = 0.0119(80{,}000)/(15.5)^2 = 3.96 \text{ ft/s}$$

Friction factors at 50,000 bbl/day and 80,000 bbl/day flow rate are

$$f_1 = 64/1190 = 0.0538$$

$$f_2 = 64/1904 = 0.0336$$

Considering a 5000 ft length of pipe, the head loss due to friction using the Darcy-Weisbach equation (3.19) is:

$$h_{L1} = 0.0538 \times (5000 \times 12/15.5) \times (2.48^2/64.4) = 19.89 \text{ ft}$$

$$h_{L2} = 0.0336 \times (5000 \times 12/15.5) \times (3.96^2/64.4) = 31.67 \text{ ft}$$

Therefore from the above it is clear in laminar flow that even though the friction factor decreases with a flow increase, the pressure drop still increases with an increase in flow rate.

For turbulent flow, when the Reynolds number $R > 4000$, the friction factor f depends not only on R but also on the internal roughness of the pipe. As the pipe roughness increases, so does the friction factor. Therefore, smooth pipes have a smaller friction factor compared with rough pipes. More correctly, friction factor depends on the relative roughness (e/D) rather than the absolute pipe roughness e.

Various correlations exist for calculating the friction factor f. These are based on experiments conducted by scientists and engineers over the last 60 years or more. A good all-purpose equation for the friction factor f in the turbulent region (i.e., where $R > 4000$) is the Colebrook-White equation:

$$1/\sqrt{f} = -2\text{Log}_{10}[(e/3.7D) + 2.51/(R\sqrt{f})] \tag{3.21}$$

where

f = Darcy friction factor, dimensionless
D = Pipe internal diameter, in.
e = Absolute pipe roughness, in.
R = Reynolds number of flow, dimensionless

In SI units, the above equation for f remains the same as long as the absolute roughness e and the pipe diameter D are both expressed in mm. All other terms in the equation are dimensionless.

It can be seen from Equation (3.21) that the calculation of f is not easy, since it appears on both sides of the equation. A trial-and-error approach needs to be used. We assume a starting value of f (say, 0.02) and substitute it in the right-hand side of Equation (3.21). This will yield a second approximation for f, which can then be used to re-calculate a better value of f, by successive iteration. Generally, three to four iterations will yield a satisfactory result for f, correct to within 0.001.

During the last two or three decades several formulas for friction factor for turbulent flow have been put forth by various researchers. All these equations attempt to simplify calculation of the friction factor compared with the Colebrook-White equation discussed above. Two such equations that are explicit equations in f, afford easy solution of friction factor compared with the implicit equation (3.21) that requires trial-and-error solution. These are called the Churchill equation and the Swamee-Jain equation and are listed in Appendix C.

In the critical zone, where the Reynolds number is between 2000 and 4000, there is no generally accepted formula for determining the friction factor. This is because the flow is unstable in this region and therefore the friction factor is indeterminate. Most users calculate the value of f based upon turbulent flow.

To make matters more complicated, the turbulent flow region ($R > 4000$) actually consists of three separate regions:

Turbulent flow in smooth pipes
Turbulent flow in fully rough pipes
Transition flow between smooth and rough pipes

For turbulent flow in smooth pipes, pipe roughness has a negligible effect on the friction factor. Therefore, the friction factor in this region depends only on the Reynolds number as follows:

$$1/\sqrt{f} = -2\mathrm{Log}_{10}[2.51/(R\sqrt{f})] \qquad (3.22)$$

For turbulent flow in fully rough pipes, the friction factor f appears to be less dependent on the Reynolds number as the latter increases in magnitude. It depends only on the pipe roughness and diameter. It can be calculated from the following equation:

$$1/\sqrt{f} = -2\text{Log}_{10}[(e/3.7D)] \tag{3.23}$$

For the transition region between turbulent flow in smooth pipes and turbulent flow in fully rough pipes, the friction factor f is calculated using the Colebrook-White equation given previously:

$$1/\sqrt{f} = -2\text{Log}_{10}[(e/3.7D) + 2.51/(R\sqrt{f})] \tag{3.24}$$

As mentioned before, in SI units the above equation for f remains the same, provided e and D are both in mm.

The friction factor equations discussed above can also be plotted on a Moody diagram as shown in Figure 3.4. Relative roughness is defined as e/D, and is simply the result of dividing the absolute pipe roughness by the pipe internal diameter. The relative roughness term is a dimensionless parameter. The Moody diagram represents the complete friction factor map for laminar and all turbulent regions of pipe flows. It is used commonly in estimating the friction factor in pipe flow. If the Moody diagram is not available, we must use trial-and-error solution of Equation (3.24)

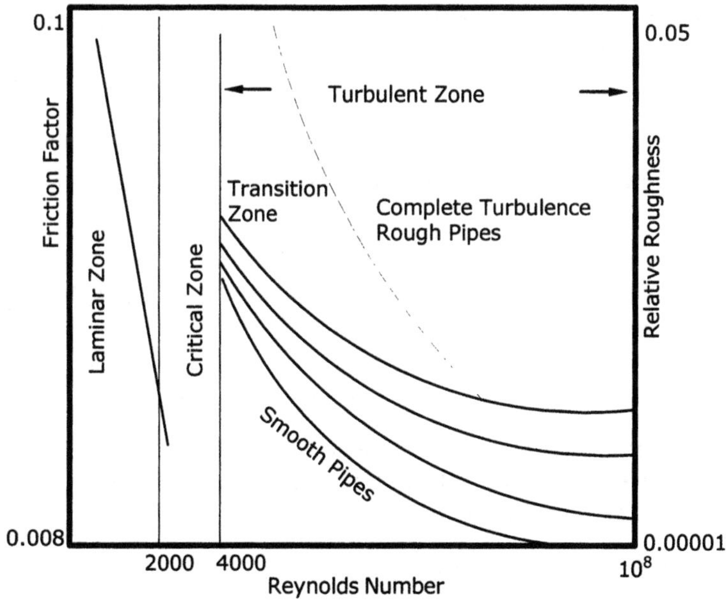

Figure 3.4 Moody diagram for friction factor.

to calculate the friction factor. To use the Moody diagram for determining the friction factor f we first calculate the Reynolds number R for the flow. Next, we find the location on the horizontal axis of Reynolds number for the value of R and draw a vertical line that intersects with the appropriate relative roughness (e/D) curve. From this point of intersection on the (e/D) curve, we read the value of the friction factor f on the vertical axis on the left.

Before leaving the discussion of the friction factor, we must mention an additional term: the Fanning friction factor. Some publications use this friction factor instead of the Darcy friction factor. The Fanning friction factor is defined as follows:

$$f_f = f_d/4 \tag{3.25}$$

where

f_f = Fanning friction factor

f_d = Darcy friction factor

Unless otherwise specified, we will use the Darcy friction factor throughout this book.

Example Problem 3.3

Water flows through a 20 in. pipe at 5700 gal/min. Calculate the friction factor using the Colebrook-White equation. Assume 0.375 in. pipe wall thickness and an absolute roughness of 0.002 in. Use a specific gravity of 1.00 and a viscosity of 1.0 cSt. What is the head loss due to friction in 2500 ft of pipe?

Solution

First we calculate the Reynolds number from Equation (3.17) as follows:

$$R = 3160 \times 5700/(19.25 \times 1.0) = 935{,}688$$

The flow is fully turbulent and the friction factor f is calculated using Equation (3.21) as follows:

$$1/\sqrt{f} = -2\mathrm{Log}_{10}[(0.002/(3.7 \times 19.25)) + 2.51/(935{,}688\sqrt{f})]$$

The above implicit equation for f must be solved by trial and error. First assume a trial value of f = 0.02. Substituting in the equation above, we get successive approximations for f as follows:

$$f = 0.0133, 0.0136, \text{ and } 0.0136$$

Therefore the solution is f = 0.0136

Using Equation (3.12),

$$\text{Velocity} = 0.4085(5700)/19.25^2 = 6.28 \text{ ft/s}$$

Using Equation (3.19), head loss due to friction is

$$h = 0.0136 \times (2500 \times 12/19.25) \times 6.28^2/64.4 = 12.98 \text{ ft}$$

3.6 Pressure Drop due to Friction

In the previous section, we introduced the Darcy-Weisbach equation:

$$h = f(L/D)(V^2/2g) \qquad (3.26)$$

where the pressure drop h is expressed in feet of liquid head and the other symbols are defined below

f = Darcy friction factor, dimensionless
L = Pipe length, ft
D = Pipe internal diameter, ft
V = Average liquid velocity, ft/s
g = Acceleration due to gravity, 32.2 ft/s^2 in English units

A more practical equation, using customary pipeline units, is given below for calculating the pressure drop in pipelines. Pressure drop due to friction per unit length of pipe, in English units, is

$$P_m = 0.0605fQ^2(Sg/D^5) \qquad (3.27)$$

and in terms of transmission factor F

$$P_m = 0.2421(Q/F)^2(Sg/D^5) \qquad (3.28)$$

where

P_m = Pressure drop due to friction, lb/in.^{2} per mile (psi/mile) of pipe length
Q = Liquid flow rate, bbl/day
f = Darcy friction factor, dimensionless
F = Transmission factor, dimensionless
Sg = Liquid specific gravity
D = Pipe internal diameter, in.

The transmission factor F is directly proportional to the volume that can be transmitted through the pipeline and therefore has an inverse relationship with the friction factor f. The transmission factor F is calculated from the following equation:

$$F = 2/\sqrt{f} \qquad (3.29)$$

Since the friction factor f ranges from 0.008 to 0.10 it can be seen from Equation (3.29) that the transmission factor F ranges from 6 to 22 approximately.

The Colebrook-White equation (3.21) can be rewritten in terms of the transmission factor F as follows:

$$F = -4\,\mathrm{Log}_{10}[(e/3.7D) + 1.255(F/R)]$$

$$\text{for turbulent flow } R > 4000 \tag{3.30}$$

Similar to the calculation of the friction factor f using Equation (3.21), the calculation of transmission factor F from Equation (3.30) will also be a trial-and-error approach. We assume a starting value of F (say 10.0) and substitute it in the right-hand side of Equation (3.30). This will yield a second approximation for F, which can then be used to recalculate a better value, by successive iteration. Generally, three or four iterations will yield a satisfactory result for F.

In SI units, the Darcy equation (in pipeline units) for the pressure drop in terms of the friction factor is represented as follows:

$$P_{km} = 6.2475 \times 10^{10} fQ^2(Sg/D^5) \tag{3.31}$$

and the corresponding equation in terms of transmission factor F is written as follows:

$$P_{km} = 24.99 \times 10^{10}(Q/F)^2(Sg/D^5) \tag{3.32}$$

where

P_{km} = Pressure drop due to friction, kPa/km
Q = Liquid flow rate, m^3/hr
f = Darcy friction factor, dimensionless
F = Transmission factor, dimensionless
Sg = Liquid specific gravity
D = Pipe internal diameter, mm

In SI units, the transmission factor F is calculated using Equation (3.30) as follows:

$$F = -4\mathrm{Log}_{10}[(e/3.7D) + 1.255(F/R)]$$

$$\text{for turbulent flow } R > 4000 \tag{3.33}$$

where

D = Pipe internal diameter, mm
e = Absolute pipe roughness, mm
R = Reynolds number of flow, dimensionless

Example Problem 3.4

Consider a 100 mile pipeline, 16 in. diameter, 0.250 in. wall thickness, transporting a liquid (specific gravity of 0.815 and viscosity of 15 cSt at 70°F) at a flow rate of 90,000 bbl/day. Calculate the friction factor and pressure drop per unit length of pipeline using the Colebrook-White equation. Assume a pipe roughness value of 0.002.

Solution

The Reynolds number is calculated first:

$$R = \frac{92.24 \times 90{,}000}{15.5 \times 15} = 35{,}706$$

Using the Colebrook-White equation (3.30), the transmission factor is

$$F = -4\ \mathrm{Log}_{10}[(0.002/(3.7 \times 15.5)) + 1.255F/35{,}706]$$

Solving above equation, by trial and error, yields

$$F = 13.21$$

To calculate the friction factor f, we use Equation (3.29) after some transposition and simplification as follows:

$$\text{Friction factor } f = 4/F^2 = 4/(13.21)^2 = 0.0229$$

The pressure drop per mile is calculated using Equation (3.28):

$$P_m = 0.2421(90{,}000/13.21)^2(0.815/15.5^5) = 10.24 \text{ psi/mile}$$

The total pressure drop in 100 miles length is then

$$\text{Total pressure drop} = 100 \times 10.24 = 1024 \text{ psi}$$

3.7 Colebrook-White Equation

In 1956 the U.S. Bureau of Mines conducted experiments and recommended a modified version of the Colebrook-White equation. The modified Colebrook-White equation yields a more conservative transmission factor F. The pressure drop calculated using the modified Colebrook-White equation is slightly higher than that calculated using the original Colebrook-White equation. This modified Colebrook-White equation, in terms of transmission factor F, is defined as follows:

$$F = -4\mathrm{Log}_{10}[(e/3.7D) + 1.4125(F/R)] \tag{3.34}$$

In SI units, the transmission factor equation above remains the same with e and D expressed in mm, other terms being dimensionless.

Comparing Equation (3.34) with Equation (3.30) or (3.33), it can be seen that the only change is in the substitution of the constant 1.4125 in place of 1.255 in the original Colebrook-White equation. Some companies use the modified Colebrook-White Equation stated in Equation (3.34).

An explicit form of an equation to calculate the friction factor was proposed by Swamee and Jain. This equation does not require trial-and-error solution like the Colebrook-White equation. It correlates very closely with the Moody diagram values. Appendix C gives a version of the Swamee-Jain equation for friction factor.

3.8 Hazen-Williams Equation

The Hazen-Williams equation is commonly used in the design of water distribution lines and in the calculation of frictional pressure drop in refined petroleum products such as gasoline and diesel. This method involves the use of the Hazen-Williams C-factor instead of pipe roughness or liquid viscosity. The pressure drop calculation using the Hazen-Williams equation takes into account flow rate, pipe diameter, and specific gravity as follows:

$$h = 4.73L(Q/C)^{1.852}/D^{4.87} \tag{3.35}$$

where

h = Head loss due to friction, ft
L = Pipe length, ft
D = Pipe internal diameter, ft
Q = Flow rate, ft^3/s
C = Hazen-Williams coefficient or C-factor, dimensionless

Typical values of the Hazen-Williams C-factor are given in Appendix A, Table A.8.

In customary pipeline units, the Hazen-Williams equation can be rewritten as follows in English units:

$$Q = 0.1482(C)(D)^{2.63}\,(P_m/Sg)^{0.54} \tag{3.36}$$

where

Q = Flow rate, bbl/day
D = Pipe internal diameter, in.
P_m = Frictional pressure drop, psi/mile
Sg = Liquid specific gravity
C = Hazen-Williams C-factor

Another form of Hazen-Williams equation, when the flow rate is in gal/min and head loss is measured in feet of liquid per thousand feet of pipe, is as follows:

$$GPM = 6.7547 \times 10^{-3}(C)(D)^{2.63}(H_L)^{0.54} \tag{3.37}$$

where

GPM = Flow rate, gal/min
H_L = Friction loss, ft of liquid per 1000 ft of pipe
Other symbols are as in Equation (3.36).

In SI units, the Hazen-Williams equation is as follows:

$$Q = 9.0379 \times 10^{-8}(C)(D)^{2.63}(P_{km}/Sg)^{0.54} \tag{3.38}$$

where

Q = Flow rate, m^3/hr
D = Pipe internal diameter, mm
P_{km} = Frictional pressure drop, kPa/km
Sg = Liquid specific gravity
C = Hazen-Williams C-factor

Historically, many empirical formulas have been used to calculate frictional pressure drop in pipelines. The Hazen-Williams equation has been widely used in the analysis of pipeline networks and water distribution systems because of its simple form and ease of use. A review of the Hazen-Williams equation shows that the pressure drop due to friction depends on the liquid specific gravity, pipe diameter, and the Hazen-Williams coefficient or C-factor.

Unlike the Colebrook-White equation, where the friction factor is calculated based on pipe roughness, pipe diameter, and the Reynolds number, which further depends on liquid specific gravity and viscosity, the Hazen-Williams C-factor appears not to take into account the liquid viscosity or pipe roughness. It could be argued that the C-factor is in fact a measure of the pipe internal roughness. However, there does not seem to be any indication of how the C-factor varies from laminar flow to turbulent flow.

We could compare the Darcy-Weisbach equation with the Hazen-Williams equation and infer that the C-factor is a function of the Darcy friction factor and Reynolds number. Based on this comparison it can be concluded that the C-factor is indeed an index of relative roughness of the pipe. It must be remembered that the Hazen-Williams equation, though convenient from the standpoint of its explicit nature, is an empirical equation, and also that it is difficult to apply to all fluids under all conditions. Nevertheless, in real-world pipelines, with sufficient field data we could determine specific C-factors for specific pipelines and fluids pumped.

Example Problem 3.5

A 3 in. (internal diameter) smooth pipeline is used to pump 100 gal/min of water. Using the Hazen-Williams equation, calculate the head loss in 3000 ft of this pipe. Assume a C-factor of 140.

Solution

Using Equation (3.37), substituting given values, we get

$$100 = 6.7547 \times 10^{-3} \times 140(3.0)^{2.63}(H_L)^{0.54}$$

Solving for the head loss we get

$$H_L = 26.6 \text{ ft per } 1000 \text{ ft}$$

Therefore head loss for 3000 ft $= 26.6 \times 3 = 79.8$ ft of water.

3.9 Shell-MIT Equation

The Shell-MIT equation, sometimes called the MIT equation, is used in the calculation of pressure drop in heavy crude oil and heated liquid pipelines. Using this method, a modified Reynolds number Rm is calculated first from the Reynolds number as follows:

$$R = 92.24(Q)/(D\nu) \tag{3.39}$$

$$Rm = R/(7742) \tag{3.40}$$

where

R = Reynolds number, dimensionless
Rm = Modified Reynolds number, dimensionless
Q = Flow rate, bbl/day
D = Pipe internal diameter, in.
ν = Kinematic viscosity, cSt

Next, depending on the flow (laminar or turbulent), the friction factor is calculated from one of the following equations:

$$f = 0.00207/Rm \qquad \text{(laminar flow)} \tag{3.41}$$

$$f = 0.0018 + 0.00662(1/Rm)^{0.355} \qquad \text{(turbulent flow)} \tag{3.42}$$

Note that this friction factor f in Equations (3.41) and (3.42) is not the same as the Darcy friction factor f discussed earlier. In fact, the friction factor f in the above equations is more like the Fanning friction factor discussed previously.

Finally, the pressure drop due to friction is calculated using the equation

$$P_m = 0.241(f\ SgQ^2)/D^5 \tag{3.43}$$

where

P_m = Frictional pressure drop, psi/mile
f = Friction factor, dimensionless
Sg = Liquid specific gravity
Q = Flow rate, bbl/day
D = Pipe internal diameter, in.

In SI units the MIT equation is expressed as follows:

$$P_m = 6.2191 \times 10^{10}(f\ SgQ^2)/D^5 \tag{3.44}$$

where

P_m = Frictional pressure drop, kPa/km
f = Friction factor, dimensionless
Sg = Liquid specific gravity
Q = Flow rate, m^3/hr
D = Pipe internal diameter, mm

Comparing Equation (3.43) with Equations (3.27) and (3.28) and recognizing the relationship between transmission factor F and Darcy friction factor f, using Equation (3.29) it is evident that the friction factor f in Equation (3.43) is not the same as the Darcy friction factor. It appears to be one-fourth the Darcy friction factor.

Example Problem 3.6

A steel pipeline of 500 mm outside diameter, 10 mm wall thickness is used to transport heavy crude oil at a flow rate of 800 m^3/hr at 100°C. Using the MIT equation calculate the friction loss per kilometer of pipe assuming an internal pipe roughness of 0.05 mm. The heavy crude oil has a specific gravity of 0.89 at 100°C and a viscosity of 120 cSt at 100°C.

Solution

From Equation (3.18)

$$\text{Reynolds number} = 353{,}678 \times 800/(120 \times 480) = 4912$$

The flow is therefore turbulent.

$$\text{Modified Reynolds number} = 4912/7742 = 0.6345$$

$$\text{Friction factor} = 0.0018 + 0.00662(1/0.6345)^{0.355} = 0.0074$$

Pressure drop from Equation (3.44) is

$$P_m = 6.2191 \times 10^{10}(0.0074 \times 0.89 \times 800 \times 800)/480^5$$

$$= 10.29\,\text{kPa/km}$$

3.10 Miller Equation

The Miller equation, also known as the Benjamin Miller formula, is used in hydraulics studies involving crude oil pipelines. This equation does not consider pipe roughness and is an empirical formula for calculating the flow rate from a given pressure drop. The equation can also be rearranged to calculate the pressure drop from a given flow rate. One of the popular versions of this equation is as follows:

$$Q = 4.06(M)(D^5P_m/Sg)^{0.5} \tag{3.45}$$

where M is defined as follows:

$$M = \text{Log}_{10}(D^3SgP_m/cp^2) + 4.35 \tag{3.46}$$

and

Q = Flow rate, bbl/day
D = Pipe internal diameter, in.
P_m = Frictional pressure drop, psi/mile
Sg = Liquid specific gravity
cp = Liquid viscosity, centipoise

In SI units, the Miller equation is as follows:

$$Q = 3.996 \times 10^{-6}(M)(D^5P_m/Sg)^{0.5} \tag{3.47}$$

where M is defined as follows:

$$M = \text{Log}_{10}(D^3SgP_m/cp^2) - 0.4965 \tag{3.48}$$

where

Q = Flow rate, m^3/hr
D = Pipe internal diameter, mm
P_m = Frictional pressure drop, kPa/km
Sg = Liquid specific gravity
cp = Liquid viscosity, centipoise

It can be seen from the above version of the Miller equation that calculating the pressure drop P_m from the flow rate Q is not straightforward. This is because the parameter M depends on the pressure drop P_m. Therefore, if we solve for P_m in terms of Q and other parameters from

Equation (3.45), we get

$$P_m = (Q/4.06M)^2(Sg/D^5) \tag{3.49}$$

where M is calculated from Equation (3.46).

To calculate P_m from a given value of flow rate Q, we use a trial-and-error approach. First, we assume a value of P_m to get a starting value of M from Equation (3.46). This value of M is then substituted in Equation (3.49) to determine a second approximation for P_m. This value of P_m will be used to generate a better value of M from Equation (3.46) which is then used to recalculate P_m. Once the successive values of P_m are within an allowable tolerance, such as 0.01 psi/mile, the iteration can be terminated and the value of pressure drop P_m has been calculated.

Example Problem 3.7

Using the Miller equation determine the pressure drop in a 14 in. diameter, 0.250 in. wall thickness, crude oil pipeline at a flow rate of 3000 gal/min. The crude oil specific gravity is 0.825 at 60°F and the viscosity is 15 cSt at 60°F.

Solution

Liquid viscosity in centipoise = 0.825 × 15 = 12.375 cP
First the parameter M is calculated from Equation (3.46) using an initial value of $P_m = 10.0$:

$$M = \text{Log}_{10}(13.5^3 \times 0.825 \times 10.0/12.375^2) + 4.35$$

$$= 6.4724$$

Using this value of M in Equation (3.49), we get

$$P_m = [3000 \times 34.2857/(4.06 \times 6.4724)]^2(0.825/13.5^5)$$

$$= 28.19 \text{ psi/mile}$$

We were quite far off in our initial estimate of P_m.
Using this value of P_m, a new value of M is calculated as

$$M = 6.9225$$

Substituting this value of M in Equation (3.49), we get

$$P_m = 24.64$$

By successive iteration, we get a final value for P_m of 25.02 psi/mile.

3.11 T. R. Aude Equation

Another pressure drop equation used in the pipeline industry that is popular among companies that transport refined petroleum products is the T. R. Aude equation, sometimes referred to simply as the Aude equation. This equation is named after the engineer who conducted experiments on pipelines in the 1950s.

The Aude equation is used in pressure drop calculations for 8 in. to 12 in. pipelines. This method requires the use of the Aude K-factor, representing pipeline efficiency. One version of this formula is given below:

$$P_m = [Q(z^{0.104})(Sg^{0.448})/(0.871(K)(D^{2.656}))]^{1.812} \tag{3.50}$$

where

P_m = Pressure drop due to friction, psi/mile
Q = Flow rate, bbl/hr
D = Pipe internal diameter, in.
Sg = Liquid specific gravity
z = Liquid viscosity, centipoise
K = T. R. Aude K-factor, usually 0.90 to 0.95

In SI units the Aude equation is as follows:

$$P_m = 8.888 \times 10^8[Q(z^{0.104})(Sg^{0.448})/(K(D^{2.656}))]^{1.812} \tag{3.51}$$

where

P_m = Frictional pressure drop, kPa/km
Sg = Liquid specific gravity
Q = Flow rate, m^3/hr
D = Pipe internal diameter, mm
z = Liquid viscosity, centipoise
K = T. R. Aude K-factor, usually 0.90 to 0.95

Since the Aude equation for pressure drop given above does not contain pipe roughness, it can be deduced that the K-factor somehow must take into account the internal condition of the pipe. As with the Hazen-Williams C-factor discussed earlier, the Aude K-factor is also an experience-based factor and must be determined by field measurement and calibration of an existing pipeline. If field data is not available, engineers usually approximate using a value such as K = 0.90 to 0.95. A higher value of K will a result in a lower pressure drop for a given flow rate or a higher a flow rate for a given pressure drop.

It must be noted that the Aude equation is based on field data collected from 6 in. and 8 in. refined products pipelines. Therefore, caution must be used when applying this formula to larger pipelines.

3.12 Minor Losses

In most long-distance pipelines, such as trunk lines, the pressure drop due to friction in the straight lengths of pipe forms the significant proportion of the total frictional pressure drop. Valves and fittings contribute very little to the total pressure drop in the entire pipeline. Hence, in such cases, pressure losses through valves, fittings, and other restrictions are generally classified as "*minor losses*". Minor losses include energy losses resulting from rapid changes in the direction or magnitude of liquid velocity in the pipeline. Thus pipe enlargements, contractions, bends, and restrictions such as check valves and gate valves are included in minor losses.

In short pipelines, such as terminal and plant piping, the pressure loss due to valves, fittings, etc., may be a substantial portion of the total pressure drop. In such cases, the term "minor losses" is a misnomer.

Therefore, in long pipelines the pressure losses through bends, elbows, valves, fittings, etc., are classified as "minor" and in most instances may be neglected without significant error. However, in shorter pipelines these losses must be included for correct engineering calculations. Experiments with water at high Reynolds numbers have shown that the minor losses vary approximately as the square of the velocity. This leads to the conclusion that minor losses can be represented by a function of the liquid velocity head or kinetic energy ($V^2/2g$).

Accordingly, the pressure drop through valves and fittings is generally expressed in terms of the liquid kinetic energy $V^2/2g$ multiplied by a head loss coefficient K. Comparing this with the Darcy-Weisbach equation for head loss in a pipe, we can see the following analogy. For a straight pipe, the head loss h is $V^2/2g$ multiplied by the factor (fL/D). Thus, the head loss coefficient for a straight pipe is fL/D.

Therefore, the pressure drop in a valve or fitting is calculated as follows:

$$h = KV^2/2g \tag{3.52}$$

where

h = Head loss due to valve or fitting, ft

K = Head loss coefficient for the valve or fitting, dimensionless

V = Velocity of liquid through valve or fitting, ft/s

g = Acceleration due to gravity, 32.2 ft/s^2 in English units

The head loss coefficient K is, for a given flow geometry, considered practically constant at high Reynolds number. K increases with pipe roughness and with lower Reynolds numbers. In general the value of K is determined mainly by the flow geometry or by the shape of the pressure loss device.

It can be seen from Equation (3.52) that K is analogous to the term (fL/D) for a straight length of pipe. Values of K are available for various types of valves and fittings in standard handbooks, such as the Crane Handbook *Flow of Fluids through Valves, Fittings, and Pipes* and *Cameron Hydraulic Data.* A table of K-factor values commonly used for valves and fittings is included in Appendix A, Table A.9.

3.12.1 Gradual Enlargement

Consider liquid flowing through a pipe of diameter D_1. If at a certain point the diameter enlarges to D_2, the energy loss that occurs due to the enlargement can be calculated as follows:

$$h = K(V_1 - V_2)^2/2g \tag{3.53}$$

where V_1 and V_2 are the velocity of the liquid in the smaller-diameter and the larger-diameter pipe respectively. The value of K depends upon the diameter ratio D_1/D_2 and the different cone angle due to the enlargement. A gradual enlargement is shown in Figure 3.5.

For a sudden enlargement $K = 1.0$ and the corresponding head loss is

$$h = (V_1 - V_2)^2/2g \tag{3.54}$$

Example Problem 3.8

Calculate the head loss due to a gradual enlargement in a pipe that flows 100 gal/min of water from a 2 in. diameter to a 3 in. diameter with an included angle of 30°. Both pipe sizes are internal diameters.

Solution

The liquid velocities in the two pipe sizes are as follows:

$$V_1 = 0.4085 \times 100/2^2 = 10.21 \text{ ft/s}$$

$$V_2 = 0.4085 \times 100/3^2 = 4.54 \text{ ft/s}$$

$$\text{Diameter ratio} = 3/2 = 1.5$$

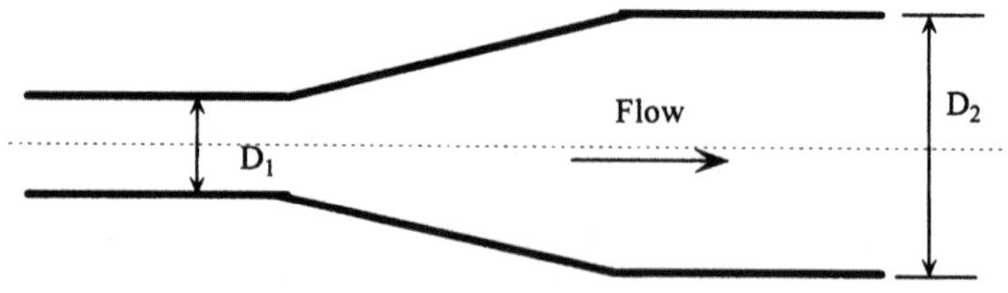

Figure 3.5 Gradual enlargement.

From charts, for diameter ratio = 1.5 and cone angle = 30° the value of K is

$$K = 0.38$$

Therefore head loss due to gradual enlargement is

$$h = 0.38 \times (10.21 - 4.54)^2/64.4 = 0.19 \text{ ft}$$

If the expansion were a sudden enlargement from 2 in. to 3 in., the head loss would be

$$h = (10.21 - 4.54)^2/64.4 = 0.50 \text{ ft}$$

3.12.2 Abrupt Contraction

For flow through an abrupt contraction, the flow from the larger pipe (diameter D_1 and velocity V_1) to a smaller pipe (diameter D_2 and velocity V_2) results in the formation of a vena contracta or throat, immediately after the diameter change. At the vena contracta, the flow area reduces to A_c with increased velocity of V_c. Subsequently the flow velocity decreases to V_2 in the smaller pipe. Thus from velocity V_1, the liquid first accelerates to velocity V_c at the vena contracta and subsequently decelerates to V_2. This is shown in Figure 3.6.

The energy loss due to the sudden contraction depends upon the ratio of the pipe diameters D_2 and D_1 and the ratio A_c/A_2. The value of the head loss coefficient K can be found using Table 3.1, where $C_c = A_c/A_2$. The ratio A_2/A_1 can be calculated from the ratio of the diameters D_2/D_1.

A pipe connected to a large storage tank represents a type of abrupt contraction. If the storage tank is a large body of liquid, we can state that this is a limiting case of the abrupt contraction. For such a square-edged pipe entrance from a large tank $A_2/A_1 = 0$. From Table 3.1, for this case $K = 0.5$ for turbulent flow.

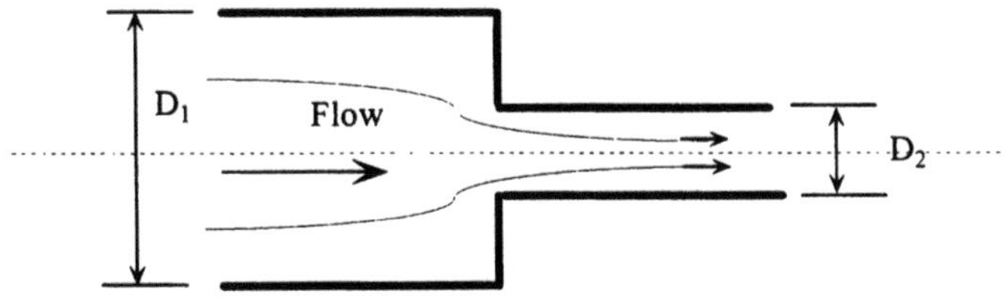

Figure 3.6 Abrupt contraction.

Table 3.1 Head Loss Coefficient K for Abrupt Contraction

A_2/A_1	C_c	K
0.0	0.617	0.50
0.1	0.624	0.46
0.2	0.632	0.41
0.3	0.643	0.36
0.4	0.659	0.30
0.5	0.681	0.24
0.6	0.712	0.18
0.7	0.755	0.12
0.8	0.813	0.06
0.9	0.892	0.02
1.0	1.000	0.00

Another type of pipe entrance from a large tank is called a re-entrant pipe entrance. If the pipe is thin-walled and the opening within the tank is located more than one pipe diameter upstream from the tank wall, the K value will be close to 0.8.

If the edges of the pipe entrance in a tank are rounded or bell-shaped, the head loss coefficient is considerably smaller. An approximate value for K for a bell-mouth entrance is 0.1.

3.12.3 Head Loss and L/D Ratio for Pipes and Fittings

We have discussed how minor losses can be accounted for using the head loss coefficient K in conjunction with the liquid velocity head. Table A.9 in Appendix A lists K values for common valves and fittings.

Referring to the Table A.9 for K values, we see that for a 16 in. gate valve, $K = 0.10$. Therefore, compared with a 16 in. straight pipe, we can write, from the Darcy-Weisbach equation:

$$\frac{fL}{D} = 0.10$$

or

$$\frac{L}{D} = \frac{0.10}{f}$$

If we assume $f = 0.0125$, we get

$$\frac{L}{D} = 8$$

This means that compared with a straight pipe 16 in. in diameter, a 16 in. gate valve has an L/D ratio of 8, which causes the same friction loss. The L/D ratio represents the equivalent length of straight pipe in terms of its diameter that will equal the pressure loss in the valve or fitting. In Table A.10 in Appendix A the L/D ratios for various valves and fittings are given.

Using the L/D ratio we can replace a 16 in. gate valve with 8 × 16 in. = 128 in. of straight 16 in. pipe. This length of pipe will have the same friction loss as the 16 in. gate valve. Thus we can use the K values or L/D ratios to calculate the friction loss in valves and fittings.

3.13 Internally Coated Pipes and Drag Reduction

In turbulent flow, pressure drop due to friction depends on the pipe roughness. Therefore, if the internal pipe surface can be made smoother, the frictional pressure drop can be reduced. Internally coating a pipeline with an epoxy coating will considerably reduce the pipe roughness compared with uncoated pipe.

For example, if the uncoated pipe has an absolute roughness of 0.002 in., coating the pipe can reduce roughness to a value as low as 0.0002 in. The friction factor f may therefore reduce from 0.02 to 0.01 depending on the flow rate, Reynolds number, etc. Since pressure drop is directly proportional to the friction factor in accordance with the Darcy-Weisbach equation, the total pressure drop in the internally coated pipeline in this example would be 50% of that in the uncoated pipeline.

Another method of reducing frictional pressure drop in a pipeline is by using drag reduction. Drag reduction is the process of reducing the pressure drop due to friction in a pipeline by continuously injecting a very small quantity (parts per million, or ppm) of a high-molecular-weight hydrocarbon, called the drag reduction agent (DRA), into the flowing liquid stream. The DRA is effective only in pipe segments between two pump stations. It degrades in performance as it flows through the pipeline for long distances. It also completely breaks up or suffers shear degradation as it passes through pump stations, meters and other restrictions. DRA works only in turbulent flow and with low-viscosity liquids. Thus, it works well with refined petroleum products (gasoline, diesel, etc.) and light crude oils. It is ineffective in heavy crude oil pipelines, particularly in laminar flow. Currently, the two leading vendors of DRA products in the United States are Baker Petrolite and Conoco-Phillips.

To determine the amount of drag reduction using DRA we proceed as follows. If the pressure drops due to friction with and without DRA are known, we can calculate the percentage drag reduction:

$$\text{Percentage drag reduction} = 100(DP_0 - DP_1)/DP_0 \qquad (3.55)$$

where

DP_0 = Friction drop in pipe segment without DRA, psi
DP_1 = Friction drop in pipe segment with DRA, psi

The above pressure drops are also referred to as untreated versus treated pressure drops. It is fairly easy to calculate the value of untreated pressure drop, using the pipe size, liquid properties, etc. The pressure drop with DRA is obtained using DRA vendor information. In most cases involving DRA, we are interested in calculating how much DRA we need to use to reduce the pipeline friction drop, and hence the pumping horsepower required. It must be noted that DRA may not be effective at the higher flow rate, if existing pump and driver limitations preclude operating at higher flow rates due to pump driver horsepower limitation.

Consider a situation where a pipeline is limited in throughput due to maximum allowable operating pressures (MAOP). Let us assume the friction drop in this MAOP limited pipeline is 800 psi at 100,000 bbl/day. We are interested in increasing pipeline flow rate to 120,000 bbl/day using DRA and we would proceed as follows:

$$\text{Flow improvement desired} = (120{,}000 - 100{,}000)/100{,}000 = 20\%$$

If we calculate the actual pressure drop in the pipeline at the increased flow rate of 120,000 bbl/day (ignoring the MAOP violation) and assume that we get the following pressure drop:

$$\text{Frictional pressure drop at } 120{,}000 \text{ bbl/day} = 1150 \text{ psi}$$

and

$$\text{Frictional pressure drop at } 100{,}000 \text{ bbl/day} = 800 \text{ psi}$$

The percentage drag reduction is then calculated from Equation (3.55) as

$$\text{Percentage drag reduction} = 100(1150 - 800)/1150 = 30.43\%$$

In the above calculation we have tried to maintain the same frictional drop (800 psi) using DRA at the higher flow rate as the initial pressure-limited case. Knowing the percentage drag reduction required, we can get the DRA vendor to tell us how much DRA will be required to achieve the 30.43% drag reduction, at the flow rate of 120,000 bbl/day. If the answer is 15 ppm of Brand X DRA, we can calculate the daily DRA requirement as follows:

$$\text{Quantity of DRA required} = (15/10^6)(120{,}000)(42) = 75.6 \text{ gal/day}$$

If DRA costs \$10 per gallon, this equates to a daily DRA cost of \$756. In this example, a 20% flow improvement requires a drag reduction of 30.43% and 15 ppm of DRA, costing \$756 per day. Of course, these are simply rough numbers, used to illustrate the DRA calculations methods. The quantity of DRA required will depend on the pipe size, liquid viscosity, flow rate, and Reynolds number, in addition to the percentage drag reduction required. Most DRA vendors will confirm that drag reduction is effective only in turbulent flow (Reynolds number > 4000) and that it does not work with heavy (high-viscosity) crude oil and other liquids.

Also, drag reduction cannot be increased indefinitely by injecting more DRA. There is a theoretical limit to the drag reduction attainable. For a certain range of flow rates, the percentage drag reduction will increase as the DRA ppm is increased. At some point, depending on the pumped liquid, flow characteristics, etc., the drag reduction levels off. No further increase in drag reduction is possible by increasing the DRA ppm. We would have reached the point of diminishing returns, in this case.

In Chapter 12 on feasibility studies and pipeline economics, we explore the subject of DRA further.

3.14 Summary

We have defined pressure and how it is measured in both a static and dynamic context. The velocity and Reynolds number calculations for pipe flow were introduced and the use of the Reynolds number in classifying liquid flow as laminar, critical, and turbulent were explained. Existing methods of calculating the pressure drop due to friction in a pipeline using the Darcy-Weisbach equation were discussed and illustrated using examples. The importance of the Moody diagram was explained. Also, the trial-and-error solutions of friction factor from the Colebrook-White equation were covered. The use of the Hazen-Williams, MIT and other pressure drop equations were discussed. Minor losses in pipelines due to valves, fittings, pipe enlargements, and pipe contractions were analyzed. The concept of drag reduction as a means of reducing frictional head loss was also introduced.

3.15 Problems

3.15.1 Calculate the average velocity and Reynolds number in a 20 in. pipeline that transports diesel fuel at a flow rate of 250,000 bbl/day. Assume 0.375 in. pipe wall thickness and the diesel fuel properties as follows: Specific gravity = 0.85, Kinematic viscosity = 5.9 cSt. What is the flow regime in this case?

3.15.2 In the above example, what is the value of the Darcy friction factor using the Colebrook-White equation? If the modified Colebrook-White equation is used, what is the difference in friction factors? Calculate the pressure drop due to friction in a 5 mile segment of this pipeline. Use a pipe roughness value of 0.002 in.

3.15.3 Using the Hazen-Williams equation with a C-factor of 125, calculate the frictional pressure drop per mile in the 20 in. pipeline described in Problem 3.15.1 above. Repeat the calculations using the Shell-MIT and Miller equations.

3.15.4 A crude oil pipeline 500 km long with 400 mm outside diameter and 8 mm wall thickness is used to transport 600 m^3/hr of product from a crude oil terminal at San José to a refinery located at La Paz. Assuming the crude oil has a specific gravity of 0.895 and viscosity of 200 SSU at 20°C, calculate the total pressure drop due to friction in the pipeline. If the MAOP of the pipeline is limited to 10 MPa, how many pumping stations will be required to transport this volume, assuming flat terrain? Use the modified Colebrook-White equation and assume a value of 0.05 mm for the absolute pipe roughness.

3.15.5 In Problem 3.15.4 the volume transported was 600 m^3/hr. It is desired to increase flow rate using DRA in the bottleneck section of the pipeline.

(a) What is the maximum throughput possible with DRA?
(b) Summarize any changes needed to the pump stations to handle the increased throughput.
(c) What options are available to further increase pipeline throughput?

4

Pipe Analysis

In this chapter we focus our attention mainly on the strength capabilities of a pipeline. We discuss different materials used to construct pipelines and how to calculate the amount of internal pressure that a given pipe can withstand. We determine the amount of internal pressure a particular size of pipe can withstand based on the pipe material, diameter, and wall thickness. Next, we establish the hydrostatic test pressure the pipeline will be subject to, such that the previously calculated internal pressure can be safely tolerated. We also discuss the volume content or line fill volume of a pipeline and how it is used in batched pipelines with multiple products.

4.1 Allowable Operating Pressure and Hydrostatic Test Pressure

To transport a liquid through a pipeline, the liquid must be under sufficient pressure so that the pressure loss due to friction and the pressure required for any elevation changes can be accommodated. The longer the pipeline and the higher the flow rate, the higher the friction drop will be, requiring a corresponding increase in liquid pressure at the beginning of the pipeline.

In gravity flow systems, flow occurs due to elevation difference without any additional pump pressure. Thus, a pipeline from a storage tank on a hill to a delivery terminus below may not need any pump pressure at the tank. However, the pipeline still needs to be designed to withstand pressure generated due to the static elevation difference.

The allowable operating pressure in a pipeline is defined as the maximum safe continuous pressure that the pipeline can be operated at. At this internal pressure the pipe material is stressed to some safe value below the yield strength of the pipe material. The stress in the pipe material consists of circumferential (or hoop) stress and longitudinal (or axial) stress. This is shown in Figure 4.1. It can be proven that the axial stress is one-half the value of the hoop stress. The hoop stress therefore controls the amount of internal pressure the pipeline can withstand. For pipelines transporting liquids, the hoop stress may be allowed to reach 72% of the pipe yield strength.

If pipe material has 60,000 psi yield strength, the safe internal operating pressure cannot exceed a value that results in a hoop stress of

$$0.72 \times 60{,}000 = 43{,}200 \text{ psi}$$

To ensure that the pipeline can be safely operated at a particular maximum allowable operating pressure (MAOP) we must test the pipeline using water, at a higher pressure.

The hydrostatic test pressure is a pressure higher than the allowable operating pressure. It is the pressure at which the pipeline is tested for a specified period of time, such as 4 hr (for aboveground piping) or 8 hr (for buried pipeline) as required by the pipeline design code or by the appropriate city or government regulations. In the United States, Department of Transportation (DOT) Code Part 195 applies. Generally, for liquid pipelines the hydrostatic test pressure is 25% higher than the MAOP. Thus, if the MAOP is 1000 psig, the pipeline will be hydrostatically tested at 1250 psig.

Calculation of internal design pressure in a pipeline is based on Barlow's equation for internal pressure in thin-walled cylindrical pipes, as discussed next.

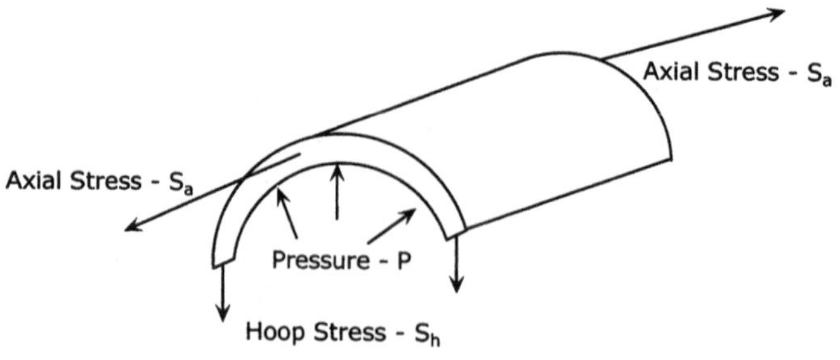

Figure 4.1 Hoop stress and axial stress in a pipe.

4.2 Barlow's Equation for Internal Pressure

The hoop stress or circumferential stress, S_h, in a thin-walled cylindrical pipe due to an internal pressure is calculated using the formula

$$S_h = PD/2t \tag{4.1}$$

where

S_h = Hoop stress, psi
P = Internal pressure, psi
D = Pipe outside diameter, in.
t = Pipe wall thickness, in.

Similarly, the axial (or longitudinal) stress, S_a, is

$$S_a = PD/4t \tag{4.2}$$

The above equations form the basis of Barlow's equation used to determine the allowable internal design pressure in a pipeline. As can be seen from Equations (4.1) and (4.2), the hoop stress is twice the longitudinal stress. The internal design pressure will therefore be based on the hoop stress (Equation 4.1).

Barlow's equation can be derived easily as follows: Consider one-half of a unit length of pipe as shown in Figure 4.1. Due to internal pressure P, the bursting force on one-half the pipe is

$$P \times D \times 1$$

where the pressure P acts on a projected area $D \times 1$. This bursting force is exactly balanced by the hoop stress S_h acting along both edges of the pipe. Therefore,

$$S_h \times t \times 1 \times 2 = P \times D \times 1$$

Solving for S_h we get

$$S_h = PD/2t$$

Equation (4.2) for axial stress S_a is derived as follows. The axial stress S_a acts on an area of cross-section of pipe represented by πDt. This is balanced by the internal pressure P acting on the internal cross-sectional area of pipe $\pi D^2/4$. Equating the two we get

$$S_a \times \pi Dt = P \times \pi D^2/4$$

Solving for S_a, we get

$$S_a = PD/4t$$

In calculating the internal design pressure in liquid pipelines, we modify Barlow's equation slightly. The internal design pressure in a pipe is calculated in English units as follows:

$$P = \frac{2T \times S \times E \times F}{D} \tag{4.3}$$

where

P = Internal pipe design pressure, psig
D = Nominal pipe outside diameter, in.
T = Nominal pipe wall thickness, in.
S = Specified minimum yield strength (SMYS) of pipe material, psig
E = Seam joint factor, 1.0 for seamless and submerged arc welded (SAW) pipes (see Table A.11 in Appendix A)
F = Design factor, usually 0.72 for liquid pipelines, except that a design factor of 0.60 is used for pipe, including risers, on a platform located off shore or on a platform in inland navigable waters, and 0.54 is used for pipe that has been subjected to cold expansion to meet the SMYS and subsequently heated, other than by welding or stress-relieving as a part of the welding, to a temperature higher than 900°F (482°C) for any period of time or to over 600°F (316°C) for more than 1 hr.

The above form of Barlow's equation may be found in Part 195 of DOT Code of Federal Regulations, Title 49 and ASME standard B31.4 for liquid pipelines. In SI units, Barlow's equation can be written as:

$$P = \frac{2T \times S \times E \times F}{D} \tag{4.4}$$

where

P = Pipe internal design pressure, kPa
D = Nominal pipe outside diameter, mm
T = Nominal pipe wall thickness, mm
S = Specified minimum yield strength (SMYS) of pipe material, kPa
E and F are defined under Equation (4.3)

In summary, Barlow's equation for internal pressure is based on calculation of the hoop stress (circumferential) in the pipe material. The hoop stress is the controlling stress within stressed pipe material, being twice the axial stress (Figure 4.1).

The strength of pipe material designated as specified minimum yield strength (SMYS) in Equations (4.3) and (4.4) depends on pipe material and grade. In the United States, steel pipeline material used in the oil and gas industry is manufactured in accordance with American Petroleum Institute (API) standards 5L and 5LX. For example, grades 5LX-42, 5LX-52, 5LX-60,

5LX-65, 5LX-70, and 5LX-80 are used commonly in pipeline applications. The numbers after 5LX above indicate the SMYS values in thousands of psi. Thus, 5LX-52 pipe has a minimum yield strength of 52,000 psi. The lowest grade of pipe material used is 5L Grade B, which has an SMYS of 35,000 psi. In addition, seamless steel pipe designated as ASTM A106 and Grade B pipe are also used for liquid pipeline systems. These have an SMYS value of 35,000 psi.

It is obvious from Barlow's equation (4.3) that, for a given pipe diameter, pipe material, and seam joint factor, the allowable internal pressure P is directly proportional to the pipe wall thickness. For example, a pipe of 16 in. diameter with a wall thickness of 0.250 in. made of 5LX-52 pipe has an allowable internal design pressure of 1170 psi calculated as follows:

$$P = (2 \times 0.250 \times 52{,}000 \times 1.0 \times 0.72)/16 = 1170\,\text{psig}$$

Therefore if the wall thickness is increased to 0.375 in. the allowable internal design pressure increases to

$$(0.375/0.250) \times 1170 = 1755\,\text{psig}$$

On the other hand, if the pipe material is changed to 5LX-70, keeping the wall thickness at 0.250 in., the new internal pressure is

$$(70{,}000/52{,}000)1170 = 1575\,\text{psig}$$

Note that we used Barlow's equation to calculate the allowable internal pressure based upon the pipe material being stressed to 72% of SMYS. In some situations more stringent city or government regulations may require that the pipe be operated at a lower pressure. Thus, instead of using a 72% factor in Equation (4.3) we may be required to use a more conservative factor (lower number) in place of $F = 0.72$. As an example, in certain areas of Los Angeles, liquid pipelines are only allowed to operate at a 66% factor instead of the 72% factor. Therefore, in the earlier example, the 16 in./0.250 in./X52 pipeline can only be operated at

$$1170(66/72) = 1073\,\text{psig}$$

As mentioned before, in order to operate a pipeline at 1170 psig, it must be hydrostatically tested at 25% higher pressure. Since 1170 psig internal pressure is based on the pipe material being stressed to 72% of SMYS, the hydrostatic test pressure will cause the hoop stress to reach

$$1.25(72) = 90\% \text{ of SMYS}$$

Generally, the hydrostatic test pressure is specified as a range of pressures, such as 90% SMYS to 95% SMYS. This is called the hydrotest pressure envelope. Therefore, in the present example the hydrotest pressure range is

1.25(1170) = 1463 psig   lower limit (90% SMYS)

(95/90)1463 = 1544 psig   higher limit (95% SMYS)

To summarize, a pipeline with an MAOP of 1170 psig needs to be hydrotested at a pressure range of 1463 psig to 1544 psig. According to the design code, the test pressure will be held for a minimum 4 hr for aboveground pipelines and 8 hr for buried pipelines.

In calculating the allowable internal pressure in older pipelines, consideration must be given to wall thickness reduction due to corrosion over the life of the pipeline. A pipeline that was installed 25 years ago with 0.250 in. wall thickness may have reduced in wall thickness to 0.200 in. or less due to corrosion. Therefore, the allowable internal pressure will have to be reduced in the ratio of the wall thickness, compared with the original design pressure.

4.3 Line Fill Volume and Batches

Frequently we need to know how much liquid is contained in a pipeline between two points along its length, such as between valves or pump stations.

For a circular pipe, we can calculate the volume of a given length of pipe by multiplying the internal cross-sectional area by the pipe length. If the pipe internal diameter is D in. and the length is L ft, the volume of this length of pipe is

$$V = 0.7854(D^2/144)L \tag{4.5}$$

where

V = Volume, ft^3

Simplifying,

$$V = 5.4542 \times 10^{-3} D^2 L \tag{4.6}$$

We will now restate this equation in terms of conventional pipeline units, such as the volume in bbl in a mile of pipe. The quantity of liquid contained in a mile of pipe, also called the line fill volume, is calculated as follows:

$$V_L = 5.129(D)^2 \tag{4.7}$$

where

V_L = Line fill volume of pipe, bbl/mile

D = Pipe internal diameter, in.

In SI units we can express the line fill volume per km of pipe as follows:

$$V_L = 7.855 \times 10^{-4}\, D^2 \tag{4.8}$$

where

V_L = Line fill volume, m^3/km

D = Pipe internal diameter, mm

Using Equation (4.7), a pipeline 100 miles long, 16 in. diameter and 0.250 in. wall thickness, has a line fill volume of

$$5.129(15.5)^2\,(100) = 123{,}224\,\text{bbl}$$

Many crude oil and refined product pipelines operate in a batched mode, in which multiple products are simultaneously pumped through the pipeline as batches. For example, 50,000 bbl of product C will enter the pipeline followed by 30,000 bbl of product B and 40,000 bbl of product A. If the line fill volume of the pipeline is 120,000 bbl, an instantaneous snapshot condition of a batched pipeline is as shown in Figure 4.2.

Example Problem 4.1

A 50 mile pipeline consists of 20 miles of pipe of 16 in. diameter and 0.375 in. wall thickness followed by 30 miles of pipe of 14 in. diameter and 0.250 in. wall thickness. Calculate the total volume contained in the 50 miles of pipeline.

Solution

Using Equation (4.7) we get, for the 16 in. pipeline

$$\text{Volume per mile} = 5.129(15.25)^2 = 1192.81\,\text{bbl/mile}$$

And for the 14 in. pipeline

$$\text{Volume per mile} = 5.129(13.5)^2 = 934.76\,\text{bbl/mile}$$

Total line fill volume is

$$20 \times 1192.81 + 30 \times 934.76 = 51{,}899\,\text{bbl}$$

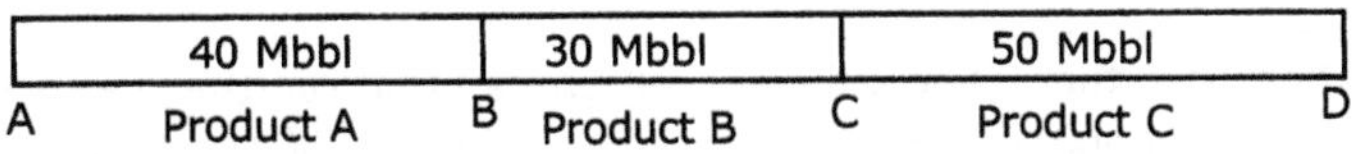

Figure 4.2 Batched pipeline.

Example Problem 4.2

A pipeline 100 km long is 500 mm outside diameter and 12 mm wall thickness. If batches of three liquids A (3000 m^3), B (5000 m^3) and C occupy the pipe, at a particular instant, calculate the interface locations of the batches, considering the origin of the pipeline to be at 0.0 km.

Solution

Using Equation (4.8) we get the line fill volume per km to be

$$V_L = 7.855 \times 10^{-4}(500 - 24)^2 = 177.9754\,m^3/km$$

The first batch A will start at 0.0 km and will end at a distance of

$$3000/177.9754 = 16.86\,km$$

The second batch B starts at 16.86 km and ends at

$$16.86 + (5000/177.9754) = 44.95\,km$$

The third batch C starts at 44.95 km and ends at 100 km.
The total volume in the pipe is

$$177.9754 \times 100 = 17{,}798\,m^3$$

Thus the volume of the third batch C is

$$17{,}798 - 3000 - 5000 = 9798\,m^3$$

It can thus be seen that line fill volume calculation is important when dealing with batched pipelines. We need to know the boundaries of each liquid batch, so that the correct liquid properties can be used to calculate pressure drops for each batch.

The total pressure drop in a batched pipeline would be calculated by adding up the individual pressure drops for each batch. Since intermixing of the batches is not desirable, batched pipelines must run in turbulent flow. In laminar flow there will be extensive mixing of the batches, which defeats the purpose of keeping each product separate; so that at the end of the pipeline each product may be diverted into a separate tank. Some intermixing will occur at the product interfaces and this contaminated liquid is generally pumped into a slop tank at the end of the pipeline and may be blended with a less critical product. The amount of contamination that occurs at the batch interface depends on the physical properties of the batched products, batch length, and Reynolds number.

4.4 Summary

In this chapter we discussed how allowable internal pressure in a pipeline is calculated depending on pipe size and material. We showed that for pipe under internal pressure the hoop stress in the pipe material will be the controlling factor. The importance of design factor in selecting pipe wall thickness was illustrated using an example. Based on Barlow's equation, the internal design pressure calculation as recommended by ASME standard B31.4 and US Code of Federal Regulation, Part 195 of the DOT was illustrated. The need for hydrostatic testing pipelines for safe operation was discussed. The line fill volume calculation was introduced and its importance in batched pipelines was shown using an example.

4.5 Problems

4.5.1 Calculate the allowable internal design pressure at 72% design factor for an 18 in. pipeline, wall thickness 0.375 in. and pipe material API 5LX-46. What is the hydrotest pressure range for this pipeline?

4.5.2 It has been determined that the design pressure for a storage tank piping system is 720 psi. If API 5L grade B pipe is used, what minimum wall thickness is required for 14 in. pipe?

4.5.3 In Problem 4.5.2, if the pressure rating were increased to ANSI 600 (1440 psi), calculate the pipe wall thickness required with 14 in. pipe if high-strength 5LX-52 pipe is used. What is the minimum hydrotest pressure for this system?

4.5.4 Determine the volume of liquid contained in a mile of 14 in. pipeline with a wall thickness of 0.281 in.

5

Pressure and Horsepower Required

In previous chapters we discussed how to calculate friction factors and pressure loss due to friction in a pipeline using various equations such as Colebrook-White, Hazen-Williams, etc. We also analyzed the internal pressure allowable in a pipe and how to determine pipe wall thickness required for a specific internal pressure and pipe material, according to design code.

In this chapter we analyze how much pressure will be required at the beginning of a pipeline to safely transport a given throughput to the pipeline terminus, taking into account the pipeline elevation profile and the pipeline terminus pressure required, in addition to friction losses. We will also calculate the pumping horsepower required and in many cases also determine how many pump stations are needed to transport the specified volume of liquid.

We also examine pipes in series and parallel and how to calculate equivalent length of pipes in series and the equivalent diameter of parallel pipes. System head curves will be introduced along with flow injection and delivery along the pipeline. Incoming and outgoing branch connections will be studied as well as pipe loops.

5.1 Total Pressure Required

The total pressure P_T required at the beginning of a pipeline to transport a given flow rate from point A to point B will depend on

Pipe diameter, wall thickness, and roughness
Pipe length
Pipeline elevation changes from A to B
Liquid specific gravity and viscosity
Flow rate

If we increase the pipe diameter, keeping all other items above constant, we know that the frictional pressure drop will decrease and hence the total pressure P_T will also decrease. Increasing pipe wall thickness or pipe roughness will cause increased frictional pressure drop and thus increase the value of P_T. On the other hand, if only the pipe length is increased, the pressure drop for the entire length of the pipeline will increase and so will the total pressure P_T.

How does the pipeline elevation profile affect P_T? If the pipeline were laid in a flat terrain, with no appreciable elevation difference between the beginning of the pipeline A and the terminus B, the total pressure P_T will not be affected. But if the elevation difference between A and B were substantial, and B was at a higher elevation than A, P_T will be higher than that for the pipeline in flat terrain.

The higher the liquid specific gravity and viscosity, the higher will be the pressure drop due to friction and hence the larger the value of P_T. Finally, increasing the flow rate will result in a higher frictional pressure drop and therefore a higher value for P_T.

In general, the total pressure required can be divided into three main components as follows:

Friction head
Elevation head
Delivery pressure at terminus

As an example, consider a pipeline from point A to point B operating at 4000 bbl/hr flow rate. If the total pressure drop due to friction in the pipeline is 800 psi, the elevation difference from point A to point B is 500 ft (uphill flow), and the minimum delivery pressure required at the terminus B is 50 psi, we can state that the pressure required at A is the sum of the three components as follows:

$$\text{Total pressure at A} = 800\,\text{psi} + 500\,\text{ft} + 50\,\text{psi}$$

If the liquid specific gravity is 0.85, then using consistent units of psi the above equation reduces to

$$\text{Total pressure} = 800 + (500)(0.85/2.31) + 50 = 1033.98\,\text{psi}$$

Of course, this assumes that there are no controlling peaks or high elevation points between point A and point B. If an intermediate point C located halfway between A and B had an elevation of 1500 ft, compared with an elevation of 100 ft at point A and 600 ft at point B, then the elevation of point C becomes a controlling factor. In this case, the calculation of total pressure required at A is a bit more complicated.

Assume that the example pipeline is 50 miles long, of uniform diameter and thickness throughout its entire length, and the flow rate is 4000 bbl/hr. Therefore, the pressure drop per mile will be calculated as a constant value for the entire pipeline from the given values as

$$800/50 = 16\,\text{psi/mile}$$

The total frictional pressure drop between point A and the peak C located 25 miles away is

$$\text{Pressure drop from A to C} = 16 \times 25 = 400\,\text{psi}$$

Since C is the midpoint of the pipeline, an identical frictional pressure drop exists between C and B as follows:

$$\text{Pressure drop from C to B} = 16 \times 25 = 400\,\text{psi}$$

Now consider the portion of the pipeline from A to C, with a frictional pressure drop of 400 psi calculated above and an elevation difference between A and C of 1500 ft − 100 ft = 1400 ft. The total pressure required at A to get over the peak at C is the sum of the friction and elevation components as follows:

$$\text{Total pressure} = 400 + (1400)(0.85/2.31) = 915.15\,\text{psi}$$

It must be noted that this pressure of 915.15 psi at A will just about get the liquid over the peak at point C with zero gauge pressure. Sometimes it is desired that the liquid at the top of the hill be at some minimum pressure above the liquid vapor pressure at the flowing temperature. If the transported liquid were liquefied petroleum gas (LPG), we would require a minimum pressure in the pipeline of 250 to 300 psi. On the other hand with crude oils and refined products with low vapor pressure, the minimum pressure required may be only 10 to 20 psi. In this example, we assume that we are dealing with low vapor pressure liquids and that therefore a minimum pressure of 10 psi is adequate at the high points in a pipeline. Our revised total pressure at A will then be

$$\text{Total pressure} = 400 + (1400)(0.85/2.31) + 10 = 925.15\,\text{psi}$$

Therefore, starting with a pressure of 925.15 psi at A will result in a pressure of 10 psi at the highest point C after accounting for the frictional pressure drop and the elevation difference between A and the peak C.

Once the liquid reaches the high point C at 10 psi, it flows downhill from point C to the terminus B at the given flow rate, being assisted by gravity. Therefore, for the section of the pipeline from C to B the elevation difference helps the flow while the friction impedes the flow. The arrival pressure at the terminus B can be calculated by considering the elevation difference and frictional pressure drop between C and B as follows:

$$\begin{aligned}\text{Delivery pressure at B} &= 10 + (1500 - 600)(0.85/2.31) - 400\\ &= -58.83\ \text{psi}\end{aligned}$$

Since the calculated pressure at B is negative, it is clear that the specified minimum delivery pressure of 50 psi at B cannot be achieved with a starting pressure of 925.15 psi at A. Therefore, to provide the required minimum delivery pressure at B, the starting pressure at A must be increased to

$$\text{Pressure at A} = 925.15 + 50 + 58.83 = 1033.98\ \text{psi}$$

The above value is incidentally the same pressure we calculated at the beginning of this section, without considering the point C. Hence the revised pressure at peak C = 10 + 50 + 58.83 = 118.83 psi.

Thus a pressure of 1033.98 psi at the beginning of the pipeline will result in a delivery pressure of 50 psi at B and clear the peak at C with a pressure of 118.83 psi. This is depicted in Figure 5.1 where

$$P_A = 1034\ \text{psi}$$
$$P_C = 119\ \text{psi}$$

and

$$P_B = 50\ \text{psi}$$

All pressures have been rounded off to the nearest whole number.

Although the higher elevation point at C appeared to be controlling, calculation showed that the pressure required at A depended more on the required delivery pressure at the terminus B. In many cases this may not be true. The pressure required will be dictated by the controlling peak and therefore the arrival pressure of the liquid at the pipeline terminus may be higher than the minimum required.

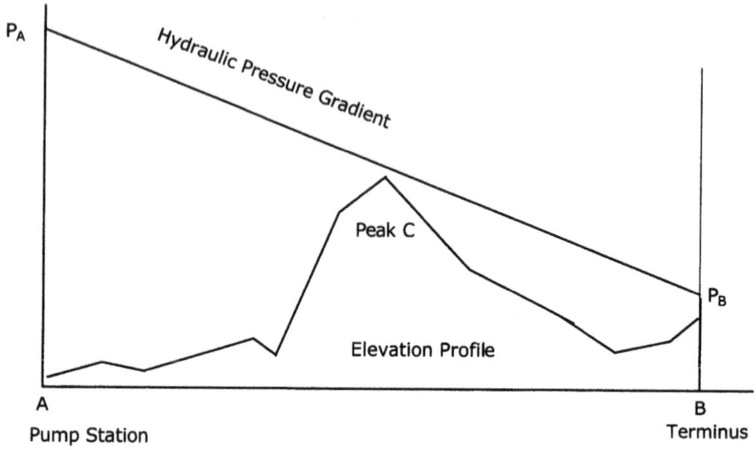

Figure 5.1 Hydraulic gradient.

Consider a second example in which the intermediate peak will be a controlling factor. Suppose the above pipeline now operates at 2200 bbl/hr and the pressure drop due to friction is calculated to be 5 psi/mile.

We will first calculate the pressure required at A, ignoring the peak elevation at C. The pressure required at A is

$$5 \times 50 + (600 - 100) \times 0.85/2.31 + 50 = 484 \text{ psi}$$

Based on this 484 psi pressure at A, the pressure at the highest point C will be calculated by deducting from 484 psi the pressure drop due to friction from A to C and adjusting for the elevation increase from A to C as follows:

$$P_C = 484 - (5 \times 25) - (1500 - 100) \times 0.85/2.31 = -156 \text{ psi}$$

This negative pressure is not acceptable, since we require a minimum positive pressure of 10 psi at the peak to prevent liquid vaporization. It is therefore clear that the pressure at A calculated above is inadequate. The controlling peak at C therefore dictates the pressure required at A. We will now calculate the revised pressure at A to maintain a positive pressure of 10 psi at the peak at C:

$$P_A = 484 + 156 + 10 = 650 \text{ psi}$$

Therefore, starting with a pressure of 650 psi at A provides the required minimum of 10 psi at the peak C. The delivery pressure at B can now be calculated as

$$P_B = 650 - (5 \times 50) - (600 - 100) \times 0.85/2.31 = 216 \text{ psi}$$

which is more than the required minimum terminus pressure of 50 psi.

This example illustrates the approach used in considering all critical elevation points along the pipeline to determine the pressure required to transport the liquid.

Next we repeat this analysis for a higher vapor pressure product.

If we were pumping a high vapor pressure liquid that requires a delivery pressure of 500 psi at the terminus, we would calculate the required pressure at A as follows:

$$\text{Pressure at A} = (6 \times 50) + \frac{(500 \times 0.65)}{2.31} + 500 = 940.69\,\text{psi}$$

where the high vapor pressure liquid (Sg = 0.65) is assumed to produce a pressure drop of 6 psi/mile for the same flow rate. The pressure at the peak will be:

$$\text{Pressure at C} = 940.69 - (6 \times 25) - (1500 - 100) \times \frac{0.65}{2.31} = 396.75\,\text{psi}$$

If the higher vapor pressure product requires a minimum pressure of 400 psi, it can be seen from above that we do not have adequate pressure at the peak C. We therefore need to increase the starting pressure at A to

$$940.69 + (400 - 396.75) = 944\ \text{psi, rounded off}$$

With this change, the delivery pressure at the terminus B will then be

$$500 + (944 - 940.69) = 503\ \text{psi}$$

This is illustrated in Figure 5.2.

5.2 Hydraulic Pressure Gradient

Generally, due to friction losses the liquid pressure in a pipeline decreases continuously from the pipe inlet to the pipe delivery terminus. If there is no elevation difference between the two ends of the pipeline and the pipe elevation profile is essentially flat, the inlet pressure at the beginning of the pipeline will decrease continuously by the friction loss at a particular flow rate. When there are elevation differences along the pipeline, the decrease in pipeline pressure along the pipeline will be due to the combined effect of pressure drop due to friction and the algebraic sum of pipeline elevations. Thus with a starting pressure of 1000 psi at the beginning of the pipeline, and assuming 15 psi/mile pressure drop due to friction in a flat

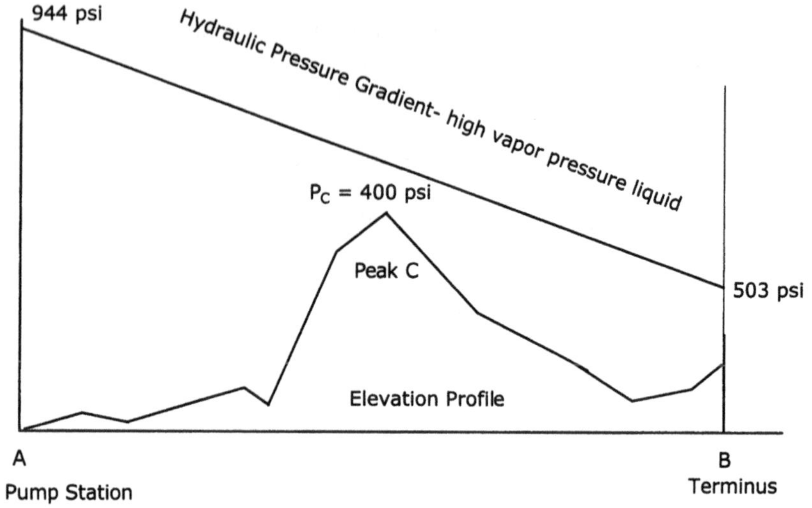

Figure 5.2 Hydraulic gradient: high vapor pressure liquid.

pipeline (no elevation difference) with constant diameter, the pressure at a distance of 20 miles from the beginning of the pipeline would drop to

$$1000 - 15 \times 20 = 700 \text{ psi}$$

If the pipeline is 60 miles long, the pressure drop due to friction in the entire line will be

$$15 \times 60 = 900 \text{ psi}$$

The pressure at the end of the pipeline will be

$$1000 - 900 = 100 \text{ psi}$$

Thus, the liquid pressure in the pipeline has uniformly dropped from 1000 psi at the beginning of the pipeline to 100 psi at the end of the 60 mile length. This pressure profile is referred to as the hydraulic pressure gradient in the pipeline. The hydraulic pressure gradient is a graphical representation of the variation in pressure along the pipeline. It is shown along with the pipeline elevation profile. Since elevation is plotted in feet, it is convenient to represent the pipeline pressures also in feet of liquid head. This is shown in Figures 5.1–5.3.

In the example discussed in Section 5.1, we calculated the pressure required at the beginning of the pipeline to be 1034 psi for pumping crude oil at a flow rate of 4000 bbl/hr. This pressure requires one pump station at the origin of the pipeline (point A). Assume now that the pipe length is 100 miles and the maximum allowable operating pressure (MAOP) is limited to 1200 psi. Suppose the total pressure required at A is calculated to be 1600 psi at a flow rate of 4000 bbl/hr. We would then require an intermediate pump station between A and B to limit the maximum pressure in the pipeline to 1200 psi. Due to the MAOP limit, the total pressure required at A will be provided in steps. The first pump station at A will provide approximately half the pressure while a second pump station located at some intermediate point provides the other half. This results in a sawtooth-like hydraulic gradient as shown in Figure 5.4. The actual discharge pressures at each pump station will be calculated considering pipeline elevations between A and B and the required minimum suction pressures at each of the two pump stations. An approximate calculation is described below, referring to Figure 5.4.

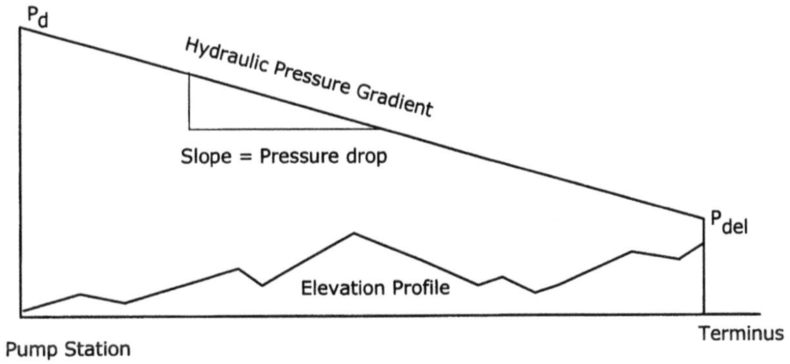

Figure 5.3 Hydraulic pressure gradient.

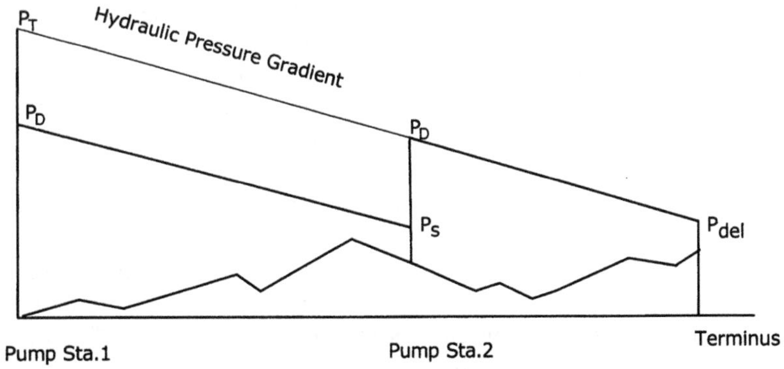

Figure 5.4 Hydraulic pressure gradient: two pump stations.

Let P_s and P_d represent the common suction and discharge pressure, respectively, for each pump station, while P_{del} is the required delivery pressure at the pipe terminus B. The total pressure P_t required at A can be written as follows.

$$P_t = P_{friction} + P_{elevation} + P_{del} \tag{5.1}$$

where

P_t = Total pressure required at A
$P_{friction}$ = Total frictional pressure drop between A and B
$P_{elevation}$ = Elevation head between A and B
P_{del} = Required delivery pressure at B

Also, from Figure 5.4, based on geometry, we can state that

$$P_t = P_d + P_d - P_s \tag{5.2}$$

Solving for P_d we get

$$P_d = (P_t + P_s)/2 \tag{5.3}$$

where

P_d = Pump station discharge pressure
P_s = Pump station suction pressure

For example, if the total pressure calculated is 1600 psi and the MAOP is 1200 psi we would need two pump stations. Considering a minimum suction pressure of 50 psi, each pump station would have a discharge pressure of

$$P_d = (1600 + 50)/2 = 825\,\text{psi} \qquad \text{using Equation (5.3)}$$

Each pump station operates at 825 psi discharge pressure and the pipeline MAOP is 1200 psi. It is clear that based on pipeline pressures alone we have the capability of increasing pipeline throughput further to fully utilize the 1200 psi MAOP at each pump station. Of course, this would correspondingly require enhancing the pumping equipment at each pump station, since more horsepower (HP) will be required at the higher flow rate. We can now estimate the increased throughput possible if we were to operate the pipeline at the 1200 psi MAOP level at each pump station.

Let us assume that the pipeline elevation difference in this example contributes 300 psi to the total pressure required. This simply represents the station elevation head between A and B converted to psi. This component of the total pressure required (P_t = 1600 psi) depends only on the pipeline elevation and liquid specific gravity and therefore does not vary with flow rate. Similarly, the delivery pressure of 50 psi at B is also independent of flow rate. We can then calculate the frictional component (which depends on flow rate) of the total pressure P_t using Equation 5.1 as follows:

$$\text{Frictional pressure drop} = 1600 - 300 - 50 = 1250\,\text{psi}$$

Assuming a pipeline length of 100 miles, the frictional pressure drop per mile of pipe is

$$P_m = 1250/100 = 12.5\ \text{psi/mile}$$

This pressure drop occurs at a flow rate of 4000 bbl/hr. From Chapter 3 on pressure drop calculations, we know that the pressure drop per mile, P_m, varies as the square of the flow rate as long as the liquid properties and pipe size do not change. Using Equation (3.28) we can write

$$P_m = K(Q)^2 \tag{5.4}$$

where

P_m = Frictional pressure drop per mile of pipe
K = A constant for this pipeline that depends on liquid properties and pipe diameter
Q = Pipeline flow rate

Note that the K value above is not the same as the head loss coefficient discussed in Chapter 3. Strictly speaking, K also includes a transmission factor F (or a friction factor f) which varies with flow rate. However, for simplicity, we will assume that K is the constant that encompasses liquid specific gravity and pipe diameter. A more rigorous approach requires an additional parameter in Equation (5.4) that would include the transmission factor F, which in turn depends on Reynolds number, pipe roughness, etc.

Therefore, using Equation (5.4) we can write for the initial flow rate of 4000 bbl/hr

$$12.5 = K(4000)^2 \tag{5.5}$$

In a similar manner, we can estimate the frictional pressure drop per mile when flow rate is increased to some value Q to fully utilize the 1200 psi MAOP of the pipeline.

Using Equation (5.3), if we allow each pump station to operate at 1200 psi discharge pressure, we can write

$$1200 = (P_t + 50)/2$$

or

$$P_t = 2400 - 50 = 2350\,\text{psi}$$

This total pressure will now consist of friction, elevation, and delivery pressure components at the higher flow rate Q. From Equation (5.1) we can write

$$2350 = P_{friction} + P_{elevation} + P_{del} \qquad \text{at the higher flow rate Q}$$

or

$$2350 = P_{friction} + 300 + 50$$

Therefore,

$$P_{friction} = 2350 - 300 - 50 = 2000 \text{ psi at the higher flow rate Q}$$

Thus the pressure drop per mile at the higher flow rate Q is

$$P_m = 2000/100 = 20\,\text{psi/mile}$$

From Equation (5.4) we can write

$$20 = K(Q)^2 \tag{5.6}$$

where Q is the unknown higher flow rate in bbl/hr.

By dividing Equation (5.6) by Equation (5.5) we get the following:

$$20/12.5 = (Q/4000)^2$$

Solving for Q we get

$$Q = 4000(20/12.5)^{1/2} = 5059.64\,\text{bbl/hr}$$

Therefore, by fully utilizing the 1200 psi MAOP of the pipeline with the two pump stations, we are able to increase the flow rate to approximately 5060 bbl/hr. As previously mentioned, this will definitely require additional pumps at both pump stations to provide the higher discharge pressure. Pumps will be discussed in Chapter 7.

In the preceding sections we have considered a pipeline of uniform diameter and wall thickness for its entire length. In reality, pipe diameter and wall thickness change, depending on the service requirement, design code, and local regulatory requirements. Pipe wall thickness may have to be increased due to differences in the specified minimum yield strength (SMYS) of pipe because a higher or lower grade of pipe was used at some locations. As mentioned previously, some cities or counties through which the pipeline traverses may require different design factors (0.66 instead of 0.72) to be used, thus necessitating a different wall thickness. If there are drastic elevation changes along the pipeline, the low elevation points may require higher wall thickness to withstand the higher pipe operating pressures. If the pipeline has intermediate flow delivery or injections, the pipe diameter may be reduced or increased for certain portions to optimize pipe use. In all these cases, we can conclude that the pressure drop due to friction will not be uniform throughout the entire pipeline length. Injections and deliveries along the pipeline and their impact on pressure required are discussed later in this chapter.

When pipe diameter and wall thickness change along a pipeline, the slope of the hydraulic gradient, as shown in Figure 5.3, will no longer be uniform. Due to varying frictional pressure drop (because of changes in pipe diameter and wall thickness), the slope of the hydraulic gradient will vary along the pipe length.

5.3 Series Piping

Pipes are said to be in series if different lengths of pipes are joined end to end with the entire flow passing through all pipes, without any branching.

Consider a pipeline consisting of two different lengths and pipe diameters joined together in series. A pipeline 1000 ft long and 16 in. in diameter connected in series with a pipeline 500 ft long and 14 in. in diameter would be an example of a series pipeline. At the connection point we will need to have a fitting, known as a reducer, that will join the 16 in. pipe with the smaller 14 in. pipe. This fitting will be a 16 in. × 14 in. reducer. The reducer causes transition in the pipe diameter smoothly from 16 in. to 14 in. We can calculate the total pressure drop through this 16 in./14 in. pipeline system by adding the individual pressure drops in the 16 in. and the 14 in. pipe segments and accounting for the pressure loss in the 16 in. × 14 in. reducer.

If two pipes of different diameters are connected together in series, we can also use the equivalent-length approach to calculate the pressure drop in the pipeline as discussed next.

A pipe is equivalent to another pipe or pipeline system when the same pressure loss due to friction occurs in the first pipe compared with that in the other pipe or pipeline system. Since the pressure drop can be caused by an infinite combination of pipe diameter and pipe length, we must specify a particular diameter to calculate the equivalent length.

Suppose a pipe A of length L_A and internal diameter D_A is connected in series with a pipe B of length L_B and internal diameter D_B. If we were to replace this two-pipe system with a single pipe of length L_E and diameter D_E, we have what is known as the equivalent length of pipe. This equivalent length of pipe may be based on one of the two diameters (D_A or D_B) or a totally different diameter D_E.

The equivalent length L_E in terms of pipe diameter D_E can be written as

$$L_E/(D_E)^5 = L_A/(D_A)^5 + L_B/(D_B)^5 \tag{5.7}$$

This formula for equivalent length is based on the premise that the total friction loss in the two-pipe system exactly equals that in the single equivalent pipe.

Equation (5.7) is based on Equation (3.28) for pressure drop, since pressure drop per unit length is inversely proportional to the fifth power of the diameter. If we refer to the diameter D_A as the basis, this equation becomes, after setting $D_E = D_A$,

$$L_E = L_A + L_B(D_A/D_B)^5 \tag{5.8}$$

Thus, we have an equivalent length L_E that will be based on diameter D_A. This length L_E of pipe diameter D_A will produce the same amount of frictional pressure drop as the two lengths L_A and L_B in series. We have thus simplified the problem by reducing it to one single pipe length of uniform diameter D_A.

It must be noted that the equivalent-length method discussed above is only approximate. Furthermore, if elevation changes are involved it becomes more complicated, unless there are no controlling elevations along the pipeline system.

An example will illustrate this concept of equivalent pipe length. Consider a 16 in. pipeline of 0.281 in. wall thickness and 20 miles long installed in series with a 14 in. pipeline of 0.250 in. wall thickness and 10 miles long. The equivalent length of this pipeline, using Equation (5.8), is

$$20 + 10 \times (16 - 0.562)^5/(14 - 0.50)^5$$

$$= 39.56 \text{ miles of } 16 \text{ in. pipe.}$$

The actual physical length of 30 miles of 16 in. and 14 in. pipes is replaced, for the purpose of pressure drop calculations with a single 16 in. pipe 39.56 miles long. Note that we have left out the pipe fitting that would connect the 16 in. pipe with the 14 in. pipe. This would be a 16 in. × 14 in. reducer which would have its own equivalent length. To be precise, we should determine the equivalent length of the reducer from Table A.10 in Appendix A and add it to the above length to obtain the total equivalent length, including the fitting.

Once an equivalent pipe length has been determined, we can calculate the pressure drop based on this pipe size.

5.4 Parallel Piping

Pipes are said to be in parallel if they are connected in such a way that the liquid flow splits into two or more separate pipes and rejoins downstream into another pipe as illustrated in Figure 5.5. In Figure 5.5 liquid flows through pipe AB until at point B part of the flow branches off into pipe BCE, while the remainder flows through pipe BDE. At point E, the flows recombine to the original value and the liquid flows through the pipe EF. Note that we are assuming that all pipes in Figure 5.5 are shown in plan view with no elevation changes.

In order to calculate the pressures and flow rates in a parallel piping system such as the one depicted in Figure 5.5, we use the following two principles of pipes in parallel:

1. Conservation of total flow
2. Common pressure loss across each parallel pipe

According to principle 1, the total flow entering each junction of pipe must equal the total flow leaving the junction, or, simply,

Total inflow = Total outflow

Thus, in Figure 5.5, all flows entering and leaving the junction B must satisfy the above principle. If the flow into the junction B is Q and the flow in branch BCE is Q_{BC} and flow in the branch BDE is Q_{BD}, we have from the above conservation of total flow:

$$Q = Q_{BC} + Q_{BD} \tag{5.9}$$

assuming Q_{BC} and Q_{BD} represent the flow out of the junction B.

The second principle of parallel pipes, defined as principle 2 above, requires that the pressure drop across the branch BCE must equal the pressure drop across the branch BDE. This is simply due to the fact that point B represents the common upstream pressure for each of these

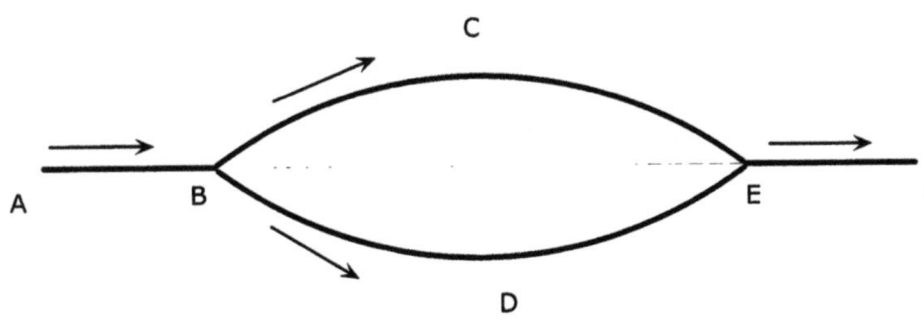

Figure 5.5 Parallel pipes.

branches, while the pressure at point E represents the common downstream pressure. Referring to these pressures as P_B and P_E, we can state

$$\text{Pressure drop in branch BCE} = P_B - P_E \tag{5.10}$$

$$\text{Pressure drop in branch BDE} = P_B - P_E \tag{5.11}$$

assuming that the flows Q_{BC} and Q_{BD} are in the direction of BCE and BDE respectively. If we had a third pipe branch between B and E, such as that shown by the dashed line BE in Figure 5.5, we can state that the common pressure drop $P_B - P_E$ would also be applicable to the third parallel pipe between B and E, as well.

We can rewrite the Equations (5.9) and (5.11) above as follows for the system with three parallel pipes:

$$Q = Q_{BC} + Q_{BD} + Q_{BE} \tag{5.12}$$

and

$$\Delta P_{BCE} = \Delta P_{BDE} = \Delta P_{BE} \tag{5.13}$$

where

ΔP = Pressure drop in respective parallel pipes.

Using the Equations (5.12) and (5.13) we can solve for flow rates and pressures in any parallel piping system. We will demonstrate the above using an example problem later in this chapter (Example Problem 5.1).

Similar to the equivalent-length concept in series piping, we can calculate an equivalent pipe diameter for pipes connected in parallel. Since each of the parallel pipes in Figure 5.5 has a common pressure drop indicated by Equation (5.13), we can replace all the parallel pipes between B and E with one single pipe of length L_E and diameter D_E such that the pressure drop through the single pipe at flow Q equals that of the individual pipes as follows:

Pressure drop in equivalent single pipe length L_E
and diameter D_E at flow rate $Q = \Delta P_{BCE}$

Assuming now that we have only the two parallel pipes BCE and BDE in Figure 5.5, ignoring the dashed line BE, we can state that

$$Q = Q_{BC} + Q_{BD}$$

and

$$\Delta P_{EQ} = \Delta P_{BCE} = \Delta P_{BDE}$$

Using Equation (3.28) the pressure ΔP_{EQ} for the equivalent pipe can be written as

$$\Delta P_{EQ} = K(L_E)(Q)^2/D_E^5$$

where K is a constant that depends on the liquid properties. The above two equations will then become

$$KL_E Q^2/D_E^5 = KL_{BC}\,Q_{BC}^2/D_{BC}^5 = K\,L_{BD}\,Q_{BD}^2/D_{BD}^5$$

Simplifying we get

$$L_E Q^2/D_E^5 = L_{BC}Q_{BC}^2/D_{BC}^5 = L_{BD}Q_{BD}^2/D_{BD}^5$$

Further simplifying the problem by setting

$$L_{BC} = L_{BD} = L_E$$

we get

$$Q^2/D_E^5 = Q_{BC}^2/D_{BC}^5 = Q_{BD}^2/D_{BD}^5$$

Substituting for Q_{BD} in terms for Q_{BC} from Equation (5.9) we get

$$Q^2/D_E^5 = Q_{BC}^2/D_{BC}^5 \tag{5.14}$$

and

$$Q_{BC}^2/D_{BC}^5 = (Q - Q_{BC})^2/D_{BD}^5 \tag{5.15}$$

From Equations (5.14) and (5.15) we can solve for the two flows Q_{BC} and Q_{BD}, and the equivalent diameter D_E, in terms of the known quantities Q, D_{BC}, and D_{BD}.

A numerical example will illustrate the above method.

Example Problem 5.1

A parallel pipe system, similar to the one shown in Figure 5.5, is located in a horizontal plane with the following data:

Flow rate Q = 2000 gal/min of water
Pipe branch BCE = 12 in. diameter, 8000 ft
Pipe branch BDE = 10 in. diameter, 6500 ft

Calculate the flow rate through each parallel pipe and the equivalent pipe diameter for a single pipe 5000 ft long between B and E to replace the two parallel pipes.

Solution

$$Q_1 + Q_2 = 2000$$

$$Q_1^2 L_1/D_1^5 = Q_2^2 L_2/D_2^5$$

where suffixes 1 and 2 refer to the two branches BCE and BDE respectively.

$$(Q_2/Q_1)^2 = (D_2/D_1)^5(L_1/L_2)$$

$$= (10/12)^5 \times (8000/6500)$$

$$Q_2/Q_1 = 0.7033$$

Solving we get

$$Q_1 = 1174 \text{ gal/min}$$

$$Q_2 = 826 \text{ gal/min}$$

The equivalent pipe diameter for a single pipe 5000 ft long is calculated as follows:

$$(2000)^2 (5000)/D_E^5 = (1174)^2 \times 8000/(12)^5$$

or

$$D_E = 13.52 \text{ in.}$$

Therefore, a 13.52 in. diameter pipe, 5000 ft long, between B and E will replace the two parallel pipes.

5.5 Transporting High Vapor Pressure Liquids

As mentioned previously, transportation of high vapor pressure liquids such as liquefied petroleum gas (LPG) requires that a certain minimum pressure be maintained throughout the pipeline. This minimum pressure must be greater than the liquid vapor pressure at the flowing temperature, otherwise liquid may vaporize causing two-phase flow in the pipeline which the pumps cannot handle. If the vapor pressure of LPG at the flowing temperature is 250 psi, the minimum pressure anywhere in the pipeline must be greater than 250 psi. Conservatively, at high elevation points or peaks along the pipeline, we must insure that more than the minimum pressure is maintained. This is illustrated in Figure 5.2. Additionally, the delivery pressure at the end of the pipeline must also satisfy the minimum pressure requirements. Thus, the delivery pressure at the pipeline terminus for LPG may be 300 psi or higher to account for any meter station and manifold piping losses at the delivery point. Also, sometimes with high vapor pressure liquids the delivery point may be a pressure vessel or a pressurized sphere maintained at 500 to 600 psi and therefore may require even higher minimum pressures compared with the vapor pressure of the liquid. Hence, both the delivery pressure and the minimum pressure must be considered when analyzing pipelines transporting high vapor pressure liquids.

5.6 Horsepower Required

So far we have examined the pressure required to transport a given amount of liquid through a pipeline system. Depending on the flow rate and MAOP of the pipeline, we may need one or more pump stations to safely transport the specified throughput. The pressure required at each pump station will generally be provided by centrifugal or positive displacement pumps. Pump operation and performance will be discussed in Chapter 7. In this section we calculate the horsepower required to pump a given volume of liquid through the pipeline regardless of the type of pumping equipment used.

5.6.1 Hydraulic Horsepower

Power required is defined as energy or work performed per unit time. In English units, energy is measured in foot pounds (ft · lb) and power is expressed in horsepower (HP). One HP is defined as 33,000 ft · lb/min or 550 ft-lb/s. In SI units, energy is measured in joules and power is measured in joules/second (watts). The larger unit kilowatt (kW) is more commonly used. One HP is equal to 0.746 kW.

To illustrate the concept of work, energy, and power required, imagine a situation that requires 150,000 gal of water to be raised 500 ft to supply the needs of a small community. If this requirement is on a 24 hr basis we can state that the work done in lifting 150,000 gal of water by 500 ft is

$$(150{,}000/7.48) \times 62.34 \times 500 = 625{,}066{,}845 \text{ ft} \cdot \text{lb}$$

where the specific weight of water is assumed to be 62.34 lb/ft^3 and 1 ft^3 = 7.48 gal. Thus we need to expend 6.25×10^8 ft · lb of energy over a 24 hr period to accomplish this task. Since 1 HP equals 33,000 ft · lb/min, the power required in this case is

$$\text{HP} = \frac{6.25 \times 10^8}{24 \times 60 \times 33{,}000} = 13.2$$

This is also known as the hydraulic horsepower (HHP), since we have not considered pumping efficiency.

As a liquid flows through a pipeline, pressure loss occurs due to friction. The pressure needed at the beginning of the pipeline to account for friction and any elevation changes is then used to calculate the amount of energy required to transport the liquid. Factoring in the time element, we get the power required to transport the liquid.

Example Problem 5.2

Consider 4000 bbl/hr being transported through a pipeline with one pump station operating at 1000 psi discharge pressure. If the pump station suction pressure is 50 psi, the pump has to produce 1000 – 50 psi, or 950 psi differential pressure to pump 4000 bbl/hr of the liquid. If the liquid specific gravity is 0.85 at flowing temperature, calculate the HP required at this flow rate.

Solution

Flow rate of liquid in lb/min is calculated as follows:

$$M = 4000\ \text{bbl/hr}\,(5.6146\ \text{ft}^3/\text{bbl})(1\ \text{hr}/60\ \text{min})(0.85)(62.34\ \text{lb/ft}^3)$$

where 62.34 is the specific weight of water in lb/ft.^3
or

$$M = 19{,}834.14\,\text{lb/min}$$

Therefore the HP required is

$$HP = \frac{(\text{lb/min})(\text{ft head})}{33{,}000}$$

or

$$HP = [(19{,}834.14)(950)(2.31)/0.85)]/33{,}000 = 1552$$

It must be noted that in the above calculation no efficiency value has been considered. In other words, we have assumed 100% pumping efficiency. Therefore, the above HP calculated is referred to as hydraulic horsepower (HHP), based on 100% efficiency.

$$HHP = 1552$$

5.6.2 Brake Horsepower

The brake horsepower takes into account the pump efficiency. If a pump efficiency of 75% is used we can calculate the brake horsepower (BHP) in the above example as follows:

$$\text{Brake horsepower} = \text{Hydraulic horsepower/Pump efficiency}$$

$$\text{BHP} = \text{HHP}/0.75 = 1552/0.75 = 2070$$

If an electric motor is used to drive the above pump, the actual motor horsepower required would be calculated as

$$\text{Motor HP} = \text{BHP/Motor efficiency}$$

Generally induction motors used for driving pumps have fairly high efficiencies, ranging from 95% to 98%. Using 98% for motor efficiency, we can calculate the motor HP required as follows:

$$\text{Motor HP} = 2070/0.98 = 2112$$

Since the closest standard-size electric motor is 2500 HP, this application will require a pump that can provide a differential pressure of 950 psi at a flow rate of 4000 bbl/hr and will be driven by a 2500 HP electric motor.

Pump companies measure pump flow rates in gal/min and pump pressures are expressed in terms of feet of liquid head. We can therefore convert the flow rate from bbl/hr to gal/min and the pump differential pressure of 950 psi can be converted to liquid head in ft.

$$4000\ \text{bbl/hr} = \frac{4000 \times 42}{60} = 2800\ \text{gal/min}$$

$$950\ \text{psi} = \frac{950 \times 2.31}{0.85} = 2582\ \text{ft}$$

The above statement for the pump requirement can then be reworded as follows: This application will use a pump that can provide a differential pressure of 2582 ft of head at 2800 gal/min and will be driven by a 2500 HP electric motor. We discuss pump performance in more detail in Chapter 7.

The formula for BHP required in terms of customary pipeline units is as follows:

$$\text{BHP} = QP/(2449E) \tag{5.16}$$

where

Q = Flow rate, bbl/hr
P = Differential pressure, psi
E = Efficiency, expressed as a decimal value less than 1.0

Two additional formulas for BHP, expressed in terms of flow rate in gal/min and pressure in psi or ft of liquid, are as follows:

$$BHP = (GPM)(H)(Spgr)/(3960E) \tag{5.17}$$

and

$$BHP = (GPM)P/(1714E) \tag{5.18}$$

where

GPM = Flow rate, gal/min
H = Differential head, ft
P = Differential pressure, psi
E = Efficiency, expressed as a decimal value less than 1.0
Spgr = Liquid specific gravity, dimensionless

In SI units, power in kW can be calculated as follows:

$$\text{Power (kW)} = \frac{Q\ H\ Spgr}{367.46(E)} \tag{5.19}$$

where

Q = Flow rate, m^3/hr
H = Differential head, m
Spgr = Liquid specific gravity
E = Efficiency, expressed as a decimal value less than 1.0

and

$$\text{Power (kW)} = \frac{Q\ P}{3600(E)} \tag{5.20}$$

where

P = Pressure, kPa
Q = Flow rate, m^3/hr
E = Efficiency, expressed as a decimal value less than 1.0

Example Problem 5.3

A water distribution system requires a pump that can produce 2500 ft head pressure to transport a flow rate of 5000 gal/min. Assuming a centrifugal pump driven by an electric motor, calculate the hydraulic HP, pump BHP, and motor HP required at 82% pump efficiency and 96% motor efficiency.

Solution

$$\text{Hydraulic HP} = \frac{5000 \times 62.34}{7.48} \times \frac{2500}{33{,}000}$$

$$HHP = 3157$$

$$\text{Pump BHP required} = \frac{3157}{0.82} = 3850\ \text{HP}$$

$$\text{Motor HP required} = \frac{3850}{0.96} = 4010\ \text{HP}$$

Example Problem 5.4

A water pipeline is used to move 320 L/s and requires a pump pressure of 750 m. Calculate the power required at 80% pump efficiency and 98% motor efficiency.

Solution

Using Equation (5.20):

$$\text{Pump power required} = \frac{320 \times 60 \times 60}{1000} \times \frac{750 \times 1.0}{367.46 \times 0.80}$$

$$= 2939\ \text{kW}$$

$$\text{Motor power required} = \frac{2939}{0.98} = 3000\ \text{kW}$$

5.7 Effect of Gravity and Viscosity

It can be seen from the previous discussions that the pump BHP is directly proportional to the specific gravity of the liquid being pumped. Therefore, if the HP for pumping water is 1000, the HP required when pumping a crude oil of specific gravity 0.85 is

$$\text{Crude oil HP} = 0.85\ (\text{Water HP}) = 0.85 \times 1000 = 850\ \text{HP}$$

Similarly, when pumping a liquid of specific gravity greater than 1.0, the HP required will be higher. This can be seen from examining Equation (5.17).

We can therefore conclude that, for the same pressure and flow rate, the HP required increases with the specific gravity of liquid pumped. The HP required is also affected by the viscosity of the liquid pumped. Consider water with a viscosity of 1.0 cSt. If a particular pump generates a head of 2500 ft at a flow rate of 3000 gal/min and has an efficiency of 85%, we can calculate the water HP using Equation (5.17) as follows:

$$\text{BHP} = \frac{3000 \times 2500 \times 1.0}{0.85 \times 3960} = 2228.16$$

If this pump is used with a liquid with a viscosity of 1000 SSU we must correct the pump head, flow rate, and efficiency values using the Hydraulic

Institute viscosity correction charts for centrifugal pumps. These are discussed in more detail in Chapter 7. The net result is that the BHP required when pumping the high-viscosity liquid will be higher than the above calculated value. It has been found that with high-viscosity liquids the pump efficiency degrades a lot faster than flow rate or head. Also, considering pipeline hydraulics, we can say that high viscosities increase the pressure required to transport a liquid and therefore increase the HP required.

The effect of viscosity and specific gravity on pump performance will be discussed in more detail in later chapters.

5.8 System Head Curves

A system head curve, also known as a system curve, for a pipeline shows the variation of pressure required with flow rate. See Figure 5.6 for a typical system head curve. As the flow rate increases, the head required increases.

Consider a pipeline of internal diameter D and length L used to transport a liquid of specific gravity Sg and viscosity ν from a pump station at A to a delivery point at B. We can calculate the pressure required at A to transport the liquid at a particular flow rate Q. By varying flow rate Q we can determine the pressure required at A for each flow rate such that a given delivery pressure at B is maintained. For each flow rate Q, we would calculate the pressure drop due to friction for the length L of the pipeline,

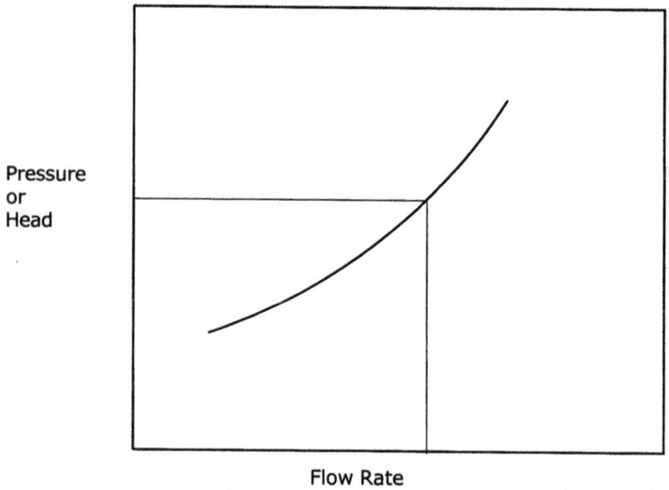

Figure 5.6 System head curve.

add the head required to account for elevation difference between A and B, and finally add the delivery pressure required at B as follows:

$$\text{Pressure at A} = \text{Frictional pressure drop} + \text{Elevation head} + \text{Delivery pressure, using Equation (5.1)}$$

Once the pressure at A is calculated for each flow rate we can plot a system head curve as shown in Figure 5.6. The vertical axis may be in feet of liquid or psi. The horizontal axis will be in units of flow such as gal/min or bbl/hr.

In Chapter 7 system head curves, along with pump head curves, will be reviewed in detail. We will see how the system head curve in conjunction with the pump head curve will determine the operating point for a particular pump-pipeline configuration.

Since the system head curve represents the pressure required to pump various flow rates through a given pipeline, we can plot a family of such curves for different liquids as shown in Figure 5.7. The higher specific gravity and viscosity of diesel fuel requires greater pressures compared with gasoline. Hence the diesel system head curve is located above that of gasoline as shown in Figure 5.7.

Note also that when there is no elevation difference involved the system head curve will start at the (0, 0) point. This means that at zero flow rate the pressure required is zero.

The shape of the system head curve varies depending upon the amount of friction head compared with elevation head. Figures 5.8 and 5.9 show two system head curves that illustrate this. In Figure 5.8 there is comparatively

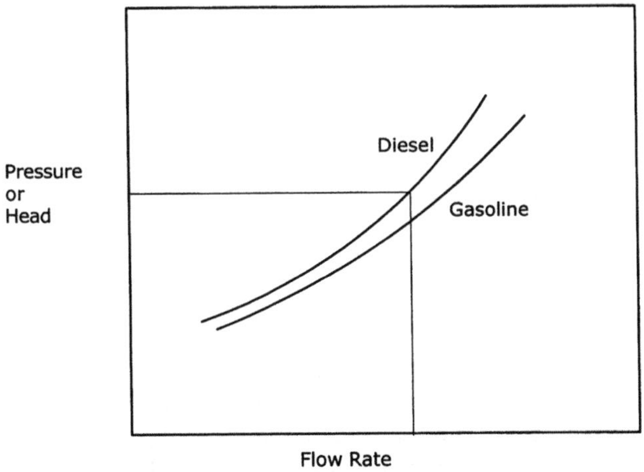

Figure 5.7 System head curve: different products.

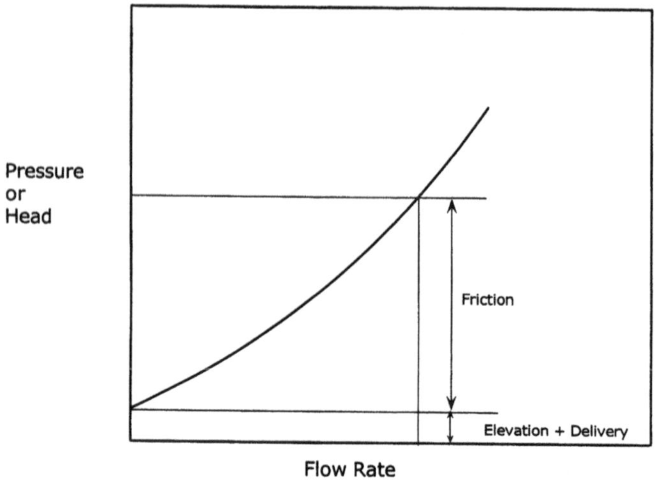

Figure 5.8 System head curve: high friction.

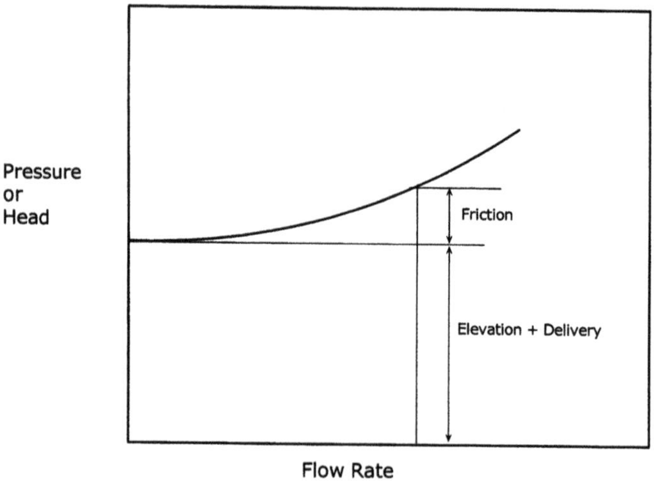

Figure 5.9 System head curve: high elevation.

less influence of the pipe elevations, most of the system head required is due to the friction in the pipe. In comparison, Figure 5.9 shows a system head curve that consists mostly of the static head due to the pipe elevations.

It can be seen from the above that the frictional head is a smaller component compared to the static elevation head.

5.9 Injections and Deliveries

In most pipelines liquid enters the pipeline at the beginning and continues to the end to be delivered at the terminus, with no deliveries or injections at any intermediate point along the pipeline. However, there are situations where liquid is delivered off the pipeline (stripping) at some intermediate location and the remainder continues to the pipeline terminus. Similarly, liquid may enter the pipeline at some intermediate location thereby adding to the existing volume in the pipeline. These are called deliveries off the pipeline and injection into the pipeline respectively. This is illustrated in Figure 5.10.

Let us analyze the pressures required in a pipeline with injection and delivery. The pipeline AB in Figure 5.10 shows 6000 bbl/hr entering the pipeline at A. At a point C, a new stream of liquid enters the pipeline at a rate of 1000 bbl/hr. Further along the pipeline, at point D, a volume of 3000 bbl/hr is stripped off the pipeline. Consequently, a resultant volume of (6000 + 1000 − 3000) or 4000 bbl/hr is delivered to the pipeline terminus at B. In order to calculate the pressures required at A for such a pipeline with injection and deliveries we proceed as follows.

First, the pipe segment between A and C that has a uniform flow of 6000 bbl/hr is analyzed. The pressure drop in AC is calculated considering the 6000 bbl/hr flow rate, pipe diameter, and liquid properties. Next the pressure drop in the pipe segment CD with a flow rate of 7000 bbl/hr is calculated taking into account the blended liquid properties by combining the incoming stream at C (6000 bbl/hr) along the main line with the injection stream (1000 bbl/hr) at C. Finally, the pressure drop in the pipe segment DB is calculated considering a volume of 4000 bbl/hr and the liquid properties in that segment, which would be the same as those of pipe segment CD. The total frictional pressure drop between A and B will be the sum of the three pressure drops calculated above. After adding any elevation head between A and B and accounting for the required delivery pressure at B we can calculate the total pressure required at point A for this pipeline system. This is illustrated in the following example.

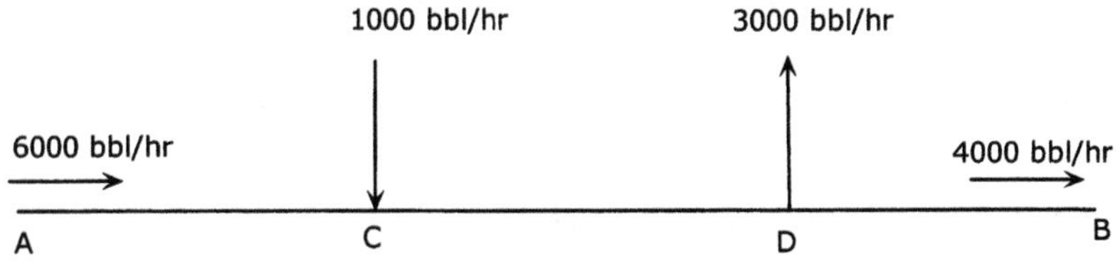

Figure 5.10 Injection and delivery.

Example Problem 5.5

In Figure 5.10 the pipeline from point A to point B is 48 miles long and is 18 in. in nominal diameter, with a 0.281 in. wall thickness. It is constructed of 5LX-65 grade steel. At A, crude oil of specific gravity 0.85 and 10 cSt viscosity enters the pipeline at a flow rate of 6000 bbl/hr. At C (milepost 22) a new stream of crude oil with a specific gravity of 0.82 and 3.5 cSt viscosity enters the pipeline at a flow rate of 1000 bbl/hr. The mixed stream then continues to point D (milepost 32) where 3000 bbl/hr is stripped off the pipeline. The remaining volume continues to the end of the pipeline at point B.

(a) Calculate the pressure required at A and the composition of the crude oil arriving at terminus B at a minimum delivery pressure of 50 psi. Assume elevations at A, C, D, and B to be 100, 150, 250, and 300 feet, respectively. Use the Colebrook-White equation for pressure drop calculations and assume a pipe roughness of 0.002 in.
(b) How much pump HP will be required to maintain this flow rate at A, assuming 50 psi pump suction pressure at A and 80% pump efficiency?
(c) If a positive displacement (PD) pump is used to inject the stream at C, what pressure and HP are required at C?

Solution

(a) The pressure drop due to friction for segment AC is calculated using Equation (3.27) as follows:

$$\text{Reynolds number} = 92.24 \times 6000 \times 24/(17.438 \times 10) = 76{,}170$$

$$\text{Friction factor} = 0.02$$

$$\text{Pressure drop} = 13.25\,\text{psi/mile}$$

$$\text{Frictional pressure drop between A and C} = 13.25 \times 22 = 291.5\text{ psi}$$

Next, we calculate the blended properties of the liquid stream after mixing two streams at point C, by blending 6000 bbl/hr of crude A (specific gravity of 0.85 and viscosity of 10 cSt) with 1000 bbl/hr of crude B (specific gravity of 0.82 and viscosity of 3.5 cSt) using Equations (2.4) and (2.21) as follows:

$$\text{Blended specific gravity at C} = 0.8457$$

$$\text{Blended viscosity at C} = 8.366\,\text{cSt}$$

For pipe segment CD we calculate the pressure drop by using above properties at a flow rate of 7000 bbl/hr.

$$\text{Reynolds number} = 92.24 \times 7000 \times 24/(17.438 \times 8.366) = 106{,}222$$

$$\text{Friction factor} = 0.0188$$

$$\text{Pressure drop} = 16.83\,\text{psi/mile}$$

$$\text{Frictional pressure drop between C and D} = 16.83 \times 10 = 168.3\,\text{psi}$$

Finally we calculate the pressure drop for pipe segment DB by using above liquid properties at a flow rate of 4000 bbl/hr:

$$\text{Reynolds number} = 92.24 \times 4000 \times 24/(17.438 \times 8.366) = 60{,}698$$

$$\text{Friction factor} = 0.021$$

$$\text{Pressure drop} = 6.13\,\text{psi/mile}$$

$$\text{Frictional pressure drop between D and B} = 6.13 \times 16 = 98.08\,\text{psi}$$

Therefore, the total frictional pressure drop between point A and point B is

$$291.5 + 168.3 + 98.08 = 557.9\,\text{psi}$$

The elevation head between A and B consists of (150 – 100) ft between A and C and (300 – 150) ft between C and B. We need to separate the total elevation head in this fashion because of differences in liquid properties in pipe segments AC and CB. Therefore, the total elevation head is

$$[(150 - 100) \times 0.85/2.31] + [(300 - 150) \times 0.8457/2.31] = 73.32\ \text{psi}$$

Adding the delivery pressure of 50 psi, the total pressure required at A is therefore

$$557.9 + 73.32 + 50 = 681.22\ \text{psi}$$

Therefore, the pressure required at A is 681.22 psi and the crude oil arriving at terminus B has a specific gravity of 0.8457 and viscosity of 8.366 cSt.

(b) The HP required at A is calculated using Equation (5.16) as follows:

$$\text{BHP} = 6000 \times (681.22 - 50)/(0.8 \times 2449) = 1933\ \text{HP}$$

(c) To calculate the injection pump requirement at point C, we must first calculate the pressure in the pipeline at point C that the PD pump has to overcome.

The pressure at C is equal to the pressure at A minus the pressure drop from A to C minus the elevation head from A to C:

$$P_C = 681.22 - 291.5 - (150 - 100) \times 0.85/2.31 = 371.3 \text{ psi}$$

The HP required for the PD pump at C is calculated using Equation (5.16) as follows:

$$\text{PD pump HP required} = (371.3 - 50) \times 1000/(0.8 \times 2449) = 164$$

assuming 50 psi suction pressure and 80% pump efficiency.

5.10 Pipe Branches

In the previous section we discussed a pipeline with an injection point and a delivery point between the pipeline inlet and pipeline terminus. These injections and deliveries may actually consist of branch pipes bringing liquid into the main line (incoming branch) and delivering liquid out of the pipeline (outgoing branch). This is illustrated in Figure 5.11.

Sometimes we are interested in sizing the branch pipes for the required flow rates and pressures. For example, in Figure 5.11 the incoming branch BR1 needs to be sized to handle 1000 bbl/hr of liquid entering the pipeline at point C with the specified pressure P_C. We have to determine the pressure required at the beginning of branch BR1 (point E) for a specified branch pipe diameter, liquid properties, etc., so that the required pressure P_C is achieved at the end of the branch at point C.

Similarly, for an outgoing branch BR2 as shown in Figure 5.11, we are interested in determining the branch pipe size such that the liquid stream flowing through the outgoing branch arrives at its destination F with a certain specified delivery pressure. We would use the starting pressure of BR2 as P_D, which represents the main line pressure at the junction with

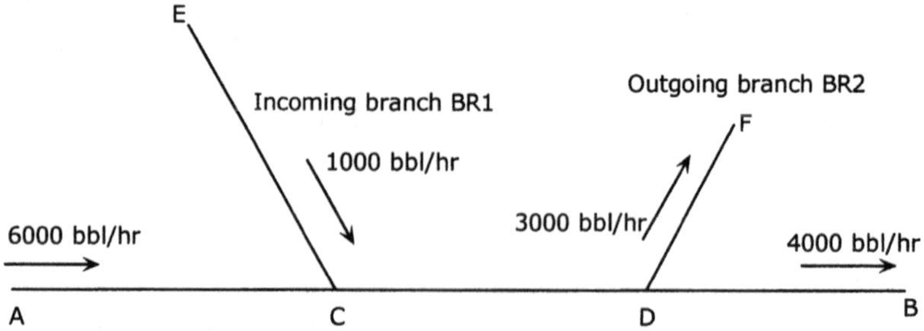

Figure 5.11 Incoming and outgoing branches.

the outgoing branch. Using the value of P_D and the required delivery pressure P_F at F we can calculate the pressure drop per mile for BR2. Hence we can select a pipe size to handle the flow rate through BR2. The following example illustrates how branch pipes are sized for injection and deliveries.

Example Problem 5.6

(a) Using the data in Example Problem 5.5 and Figure 5.11, determine the branch pipe sizes required for branch BR1 for the injection rate of 1000 bbl/hr and an MAOP of 600 psi. Assume pipe branch BR1 to be 2.5 miles long and an essentially flat elevation profile.
(b) Calculate the pipe size for the outgoing branch BR2 to handle the delivery of 3000 bbl/hr at D and pressure of 75 psi at F. The branch BR2 is 4 miles long.
(c) What HP is required at beginning of branch BR1?

Solution

(a) Assume a 6 in. pipe for branch BR1 and calculate the pressure drop in the 2.5 miles of pipe as follows: Using given liquid properties we calculate:

Pressure drop = 65.817 psi/mile for 1000 bbl/hr in a 6 in. pipe

Pressure required at E to match junction pressure of 371.3 at C is

$$P_E = 65.817 \times 2.5 + 371.3 = 535.84 \text{ psi}$$

Since this is less than the MAOP of 600 psi for BR1, a 6 in. line is adequate.

(b) With a junction pressure of 166.39 psi available at D, a 10 in. pipe is inadequate to carry 3000 bbl/hr of crude through the 4 miles of branch pipe BR2 and provide a 75 psi delivery pressure at F. Next, assuming a 12 in. pipe for BR2, 3000 bbl/hr flow rate, specific gravity of 0.8457 and viscosity of 8.366 cSt, we get

Pressure drop in BR2 = 20.07 psi/mile

Therefore, we get a delivery pressure of

$$P_F = 166.39 - 20.07 \times 4 = 86.11 \text{ psi}$$

which is higher than the minimum 50 psi delivery pressure required. Therefore, branch pipe BR2 must be at least 12 in. nominal diameter.

(c) The HP required at beginning of branch BR1 is calculated to be as follows using Equation (5.16):

$$HP = (535.84 - 50) \times 1000/(2449 \times 0.8) = 248$$

based on 80% efficiency and 50 psi suction pressure at E.

5.11 Pipe Loops

A pipe loop is a length of parallel pipe installed between two points on a main pipeline as shown in Figure 5.12. We discussed parallel pipes and equivalent diameter earlier in this chapter. In this section we will discuss how looping an existing pipeline will reduce the pressure drop due to friction and thus require less pumping HP.

The purpose of the pipe loop is to split the flow through a parallel segment of the pipeline between the two locations, resulting in a reduced pressure drop in that segment of the pipeline.

Consider pipeline from A to B with a loop installed from C to D as shown in Figure 5.12. The flow rate between A and C is 6000 gal/min. At C where the loop is installed the flow rate of 6000 gal/min is partially diverted to the loop section with the remainder going through the mainline portion CD. If we assume that the diameter of the loop is the same as that of the main pipe, this will cause 3000 gal/min flow through the loop and an equal amount through the mainline. Assume that the section CD of the mainline prior to looping with the full flow of 6000 gal/min flowing through it had a resulting pressure drop of 25 psi/mile. With the loop installed, section CD has half the flow and therefore approximately one-fourth of the pressure drop (since the pressure drop varies as the square of the flow rate) or 6.25 psi/mile, based on Equation (3.27). If the length CD is 10 miles, the total frictional pressure drop without the loop will be 250 psi. With the pipe loop the pressure drop will be 62.5 psi, which is a significant reduction. Therefore if a pipeline section is bottlenecked due to maximum allowable operating pressure, we can reduce the overall pressure profile by installing a loop in that pipe segment.

To illustrate this concept further, consider the following example problem.

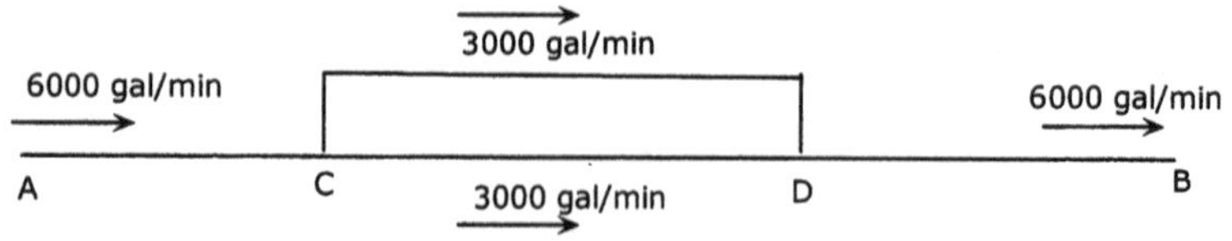

Figure 5.12 Pipe loop.

Example Problem 5.7

A 16 in. crude oil pipeline (0.250 in. wall thickness) is 30 miles long from point A to point B. The flow rate at the inlet A is 4000 bbl/hr. The crude oil properties are specific gravity of 0.85 and viscosity of 10 cSt at a flowing temperature of 70°F.

(a) Calculate the pressure required at A without any pipe loop. Assume 50 psi delivery pressure at the terminus B and a flat pipeline elevation profile.
(b) If a 10 mile portion CD, starting at milepost 10, is looped with an identical 16 in. pipeline, calculate the reduced pressure at A.
(c) What is the difference in pump HP required at A between cases (a) and (b) above? Assume 80% pump efficiency and 25 psi pump suction pressure.

Solution

Using the Colebrook-White equation and assuming a pipe roughness of 0.002 in., we calculate the pressure drop per unit length using Equation (3.27) as follows:

$$\text{Reynolds number} = 92.24 \times 4000 \times 24/(15.5 \times 10) = 57{,}129$$

$$\text{Friction factor } f = 0.0213$$

$$\text{Pressure drop} = 11.28 \text{ psi/mile}$$

(a) Therefore the total pressure required at $A = 30 \times 11.28 + 50 = 388.40$ psi without any pipe loop. With 10 miles of pipe loop the flow rate through the loop is 2000 bbl/hr and the revised pressure drop at the reduced flow is calculated as

$$\text{Pressure drop} = 3.28 \text{ psi/mile}$$

The pressure drop in the 10 mile loop $= 10 \times 3.28 = 32.8$ psi
The pressure drop in the 10 mile section $AC = 10 \times 11.28 = 112.8$ psi
The pressure drop in the last 10 mile section $DB = 10 \times 11.28 = 112.8$ psi
Therefore

(b) Total pressure required at $A = 112.8 + 32.8 + 112.8 + 50 = 308.40$ psi with the pipe loop.

Therefore, using the 10 mile loop causes a reduction of 80 psi in the pressure required at point A.

(c) HP required without the loop is

$$HP = 4000 \times (388.40 - 25)/(2449 \times 0.8) = 742$$

HP required considering the loop is

$$HP = 4000 \times (308.40 - 25)/(2449 \times 0.8) = 579$$

Thus, installing the pipe loop results in a reduction in pumping HP of

$$(742 - 579)/742 = 22\%$$

What would be the impact if the looped section of pipe were smaller than the mainline? If we install a smaller pipe in parallel, the flow will still be split through the loop section but will not be equally divided. The smaller pipe will carry less flow rate than the larger mainline pipe, such that the pressure drop through the mainline from point C to point D will exactly equal the pressure drop through the pipe loop between C and D, since both the mainline pipe and the pipe loop have common pressures at the junction C and D.

To illustrate this, consider an 8 in. pipe looped with the 16 in. mainline. Assume Q_8 represents the flow rate through the 8 in. loop and Q_{16} the flow rate through the 16 in. mainline portion. We can write

$$Q_8 + Q_{16} = \text{Total flow} = 4000 \text{ bbl/hr} \tag{5.21}$$

From Equation (3.27), we know that the frictional pressure drop in a pipe is directly proportional to the square of the flow rate and inversely proportional to the fifth power of the diameter. Therefore, we can write

$$\text{Pressure drop in 8 in. pipe} = K(Q_8)^2/(8.625 - 0.5)^5$$

considering 0.250 in. wall thickness and where K is a constant,
or

$$\Delta P_8 = K(Q_8)^2/35{,}409 \tag{5.22}$$

Similarly, for the 16 in. pipe

$$\Delta P_{16} = K(Q_{16})^2/894{,}661 \tag{5.23}$$

Dividing Equation (5.22) by Equation (5.23), we get

$$\Delta P_8/\Delta P_{16} = 25.27(Q_8/Q_{16})^2$$

Since ΔP_8 and ΔP_{16} are equal in a loop system, the above equation reduces to

$$Q_{16} = 5.03 \times Q_8 \tag{5.24}$$

Solving Equation (5.21) and Equation (5.24) simultaneously, we get

$$Q_8 = 4000/6.03 = 664 \text{ bbl/hr} \qquad \text{flow through the 8 in. loop}$$

and

$Q_{16} = 4000 - 664 = 3336$ bbl/hr flow through the 16 in. mainline portion

We can now calculate the pressure drop and the total pressure required at A as we did in Example Problem 5.6 earlier.

Since the 10 mile looped portion of the 16 in. pipe has a flow rate of 3336 bbl/hr we calculate the pressure drop in this pipe first:

Reynolds number = 92.24 × 3336 × 24/(15.5 × 10) = 47,646

Friction factor f = 0.0216

and

Pressure drop = 7.96 psi/mile

Therefore, the pressure drop in the loop section

= 7.96 × 10 = 79.6 psi

The pressure drops in section AC and section DB remain the same as before. The total pressure required at A becomes

Pressure at A = 112.8 + 79.6 + 112.8 + 50 = 355.2 psi

Compare this with 388.4 psi (without loop) and 308.4 psi (with 16 in. loop).

We could also calculate the above using the equivalent-diameter approach discussed earlier in this chapter. We will calculate the equivalent diameter of a 10 mile pipe that will replace the 8 in./16 in. loop.

Recalling our discussions of parallel pipes and equivalent diameter in Section 5.4 and using Equations (5.14) and (5.15), we calculate the equivalent diameter D_E as follows (after some simplifications):

$$D_E^{2.5} = (8.125)^{2.5} + (15.5)^{2.5}$$
$$D_E = 16.67 \text{ in.}$$

Therefore, we can say that the two parallel pipes of diameters 8 in. and 16 in. together equal a single pipe of 16.67 in. internal diameter. To show that the equivalent-diameter concept is fairly accurate, we will calculate the pressure drop in this equivalent pipe and compare it with that for the two parallel pipes.

Reynolds number = 92.24 × 4000 × 24/(16.67 × 10) = 53,120

Friction Factor f = 0.0211

and

Pressure drop = 7.77 psi/mile

Compare this with 7.96 psi/mile we calculated earlier. It can be seen that the equivalent-diameter method is within 2% of the more exact method and is therefore accurate for most practical purposes.

A practice problem with dissimilar pipe sizes as discussed above is included at the end of this chapter as an exercise for the reader.

5.12 Summary

In this chapter we extended the pressure drop concept developed in Chapter 3 to calculate the total pressure required to transport liquid through a pipeline taking into account the elevation profile of the pipeline and required delivery pressure at the terminus. We analyzed pipes in series and parallel and introduced the concept of equivalent pipe length and equivalent pipe diameter. System head curve calculations were discussed to compare pressure required at various flow rates through a pipe segment. Flow injection and delivery in pipelines were studied and the impact on the hydraulic gradient discussed. We also sized branch piping connections and discussed the advantage of looping a pipeline to reduce overall pressure drop and pumping horsepower. For a given pipeline system, the hydraulic horsepower, brake horsepower, and motor horsepower calculations were illustrated using examples. A more comprehensive analysis of pumps and HP required will be covered in Chapter 7.

5.13 Problems

5.13.1 A pipeline 50 miles long, 16 in. outside diameter, 0.250 in. wall thickness is constructed of API 5LX-65 material. It is used to transport diesel and other refined products from the refinery at Carson to a storage tank at Compton. During phase I, a flow rate of 5000 bbl/hr of diesel fuel is to be transported with one pump station located at Carson. The required delivery pressure at Compton is 50 psi. Assume a generally rolling pipeline elevation profile without any critical peaks along the pipeline. The elevation at Carson is 100 ft, and the storage tank at Compton is located on top of a hill at an elevation of 350 ft.

(a) Using diesel with specific gravity of 0.85 and a viscosity of 5.5 cSt at a flowing temperature of 70°F, calculate the total pressure required at Carson to transport 5000 bbl/hr of diesel on a continuous basis. Use the Hazen-Williams equation with a C-factor of 125. The Carson pump suction pressure is 30 psi and the pump is driven by a constant-speed electric motor.

(b) Determine the BHP and motor HP required at Carson assuming 82% pump efficiency and 96% motor efficiency.

(c) What size electric motor would be required at Carson?

(d) Assuming a maximum allowable operating pressure of 1400 psi for the pipeline, how much additional throughput can be achieved in phase II if the pumps are modified at Carson?

5.13.2 A water pipeline is being built to transport 2300 m^3/hr from a storage tank at Lyon (elevation 500 m) to a distribution facility 50 km away at Fenner (elevation 850 m). What size pipe will be required to limit velocities to 3 m/s and allowable pipeline pressure of 5.5 MPa? No intermediate pump stations are to be used. Delivery pressure at Fenner is 0.3 MPa. Use the Hazen-Williams equation with a C-factor of 110.

5.13.3 In Problem 5.13.1, what throughputs can be achieved when pumping gasoline alone? Use a specific gravity of 0.74 and viscosity of 0.65 for gasoline at the flowing temperature. Compare the pump head requirements when pumping diesel versus gasoline. Use the Hazen-Williams equation with a C-factor of 145.

5.13.4 Calculate the BHP requirements for the above problem for both diesel and gasoline movements.

5.13.5 Consider a loop pipeline similar to that in Example Problem 5.7. Instead of the loop being 16 in. consider a smaller 10 in. pipe installed in parallel for the middle 10 mile section. How does the pressure at A change compared with the no-loop case?

6

Multi-Pump Station Pipelines

In Chapter 5 we calculated the total pressure required to pump a liquid through a pipeline from point A to point B at a specified flow rate. Three components of the total pressure required (friction head, elevation head, and delivery pressure) were analyzed. Depending on the maximum allowable operating pressure (MAOP) of the pipeline we concluded that one or more pump stations may be required to handle the throughput.

In this chapter we will discuss pipeline hydraulics for multiple pump stations. Hydraulic balance and how to determine the intermediate booster pump stations will be explained. To utilize pipe material efficiently we will explore telescoping pipe wall thickness and grade tapering. Also covered will be slack line flow and a more detailed analysis of batched pipeline hydraulics.

6.1 Hydraulic Balance and Pump Stations Required

Suppose calculations indicate that at a flow rate of 5000 gal/min, a 100 mile pipeline requires a pressure of 2000 psi at the beginning of the pipeline. This 2000 psi pressure may be provided in two steps of 1000 psi each or three steps of approximately 670 psi each. In fact, due to the internal pressure limit of the pipe, we may not be able to provide one pump station at the beginning of the pipeline, operating at 2000 psi. Most pipelines have an internal pressure limit of 1000 to 1440 psi based on pipe wall thickness, grade of steel, etc., as we found in Chapter 4. Therefore, in long pipelines the total pressure required to pump the liquid is provided in two or more stages by installing intermediate booster pumps along the pipeline.

In the example case with a 2000 psi requirement and 1400 psi pipeline MAOP, we would provide this pressure as follows. The pump station at the start of the pipeline will provide a discharge pressure of 1000 psi, which will be consumed by friction loss in the pipeline and at some point (roughly halfway) along the pipeline the pressure will drop to zero. At this location we boost the liquid pressure to 1000 psi using an intermediate booster pump station. We have assumed that the pipeline is essentially on a flat elevation profile.

This pressure of 1000 psi will be sufficient to take care of the friction loss in the second half of the pipeline length. The liquid pressure will reduce to zero at the end of the pipeline. Since the liquid pressure at any point along the pipeline must be above the vapor pressure of the liquid at the flowing temperature, and the intermediate pumps require certain minimum suction pressure, we cannot allow the pressure at any point to drop to zero. Accordingly, we will locate the second pump station at a point where the pressure has dropped to a suitable suction pressure, such as 50 psi. The minimum suction pressure required is also dictated by the particular pump and may have to be higher than 50 psi, to account for any restrictions and suction piping losses at the pump station. For the present, we will assume 50 psi suction pressure is adequate for each pump station. Hence, starting with a discharge pressure of 1050 psi (1000 + 50) we will locate the second pump station (intermediate booster pump) along the pipeline where the pressure has dropped to 50 psi. This pump station will then boost the liquid pressure back up to 1050 psi and will deliver the liquid to the pipeline terminus at 50 psi. Thus each pump station provides 1000 psi differential pressure (discharge pressure minus suction pressure) to the liquid, together matching the total pressure requirement of 2000 psi at 5000 gal/min flow rate.

Note that in the above analysis we ignored pipeline elevations and assumed that the pipeline profile is essentially flat. With elevations taken into account, the location of the intermediate booster pump will be different from that of a pipeline along a flat terrain.

Hydraulic balance is when each pump station supplies the same amount of energy to the liquid. Ideally pump stations will be located at hydraulic centers. This will result in the same horsepower (HP) being added to the liquid at each pump station. For a single flow rate at the inlet of the pipeline (no intermediate injections or deliveries), the hydraulic centers will also result in the same discharge pressures at each pump station. Due to topographic conditions it may not be possible to locate the intermediate pump station at the locations desired for hydraulic balance. For example, calculations may show that three pump stations are required to handle the flow rate and that the two intermediate pump stations are to be located at milepost 50 and milepost 85. The location of milepost 50, when investigated in the field, may be found to be in the middle of a swamp or a river. Hence we will have to relocate the pump station to a more suitable location after field investigation. If the revised location of the second pump station were at milepost 52, then obviously hydraulic balance would no longer be valid. Recalculations of the hydraulics with the newly selected pump station locations will show hydraulic imbalance and all pump stations will not be operating at the same discharge pressure or providing the same amount of HP to the liquid at each pump station. However, while it is desirable to have all pump stations balanced, it may not be practical. Balanced pump station locations afford the advantage of using identical pumps and motors and the convenience of maintaining a common set of spare parts (pump rotating elements, mechanical seal, etc.) at a central operating district location.

In Chapter 5 we discussed how the location of an intermediate pump station can be calculated from given data on pump station suction pressure, discharge pressure, etc. In Section 5.2 we presented a formula to calculate the discharge pressure for a pipeline system with two pump stations given the total pressure required for a particular flow rate. We will expand on that discussion by presenting a method to calculate the pump station pressures for hydraulic balance.

Figure 6.1 shows a pipeline with varying elevation profile but no significant controlling peaks along the pipeline. The total pressure P_T was calculated for the given flow rate and liquid properties. The hydraulic

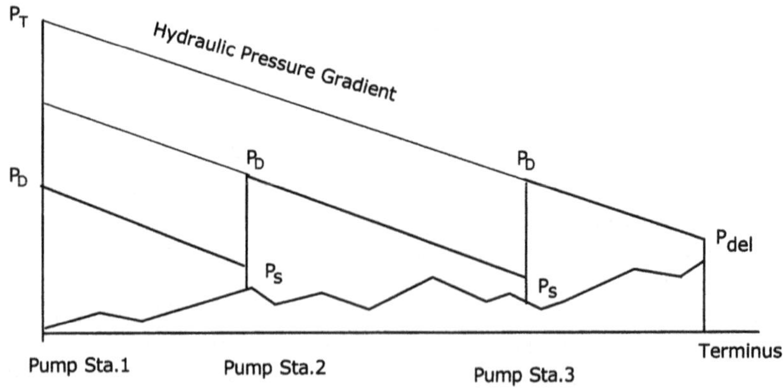

Figure 6.1 Hydraulic gradient: multiple pump stations.

gradient with one pump station at the total pressure P_T is as shown. Since P_T may be higher than the pipe MAOP, we will assume that three pump stations are required to provide the pressures needed within MAOP limits. Each pump station will discharge at pressure P_D. If P_S represents the pump station suction pressure and P_{del} the delivery pressure at the pipeline terminus, we can state the following, using geometry:

$$P_D + (P_D - P_S) + (P_D - P_S) = P_T \tag{6.1}$$

Since the above is based on one origin pump station and two intermediate pump stations, we can extend the above equation for N pump stations as follows:

$$P_D + (N - 1) \times (P_D - P_S) = P_T \tag{6.2}$$

Solving for N we get

$$N = (P_T - P_S)/(P_D - P_S) \tag{6.3}$$

Equation (6.3) is used to estimate the number of pump stations required for hydraulic balance given the discharge pressure limit P_D at each pump station. Solving Equation (6.3) for the common pump station discharge pressure,

$$P_D = (P_T - P_S)/N + P_S \tag{6.4}$$

As an example, if the total pressure calculated is 2000 psi and the suction pressure is 25 psi, the number of pump stations required with 1050 psi discharge pressure is

$$N = (2000 - 50)/(1050 - 50) = 1.95$$

Rounding up to the nearest whole number, we can conclude that two pump stations are required. Using Equation (6.4), the discharge pressure P_D at which each pump station will operate is:

$$P_D = (2000 - 50)/2 + 50 = 1025\,\text{psi}$$

Once we have calculated the discharge pressure required for hydraulic balance, as above, a graphical method can be used to locate the pump stations along the pipeline profile. First the pipeline profile (milepost versus elevation) is plotted and the hydraulic gradient superimposed upon it by drawing the sloped line starting at P_T at A and ending at P_{del} at D, as shown in Figure 6.1. Note that the pressure must be converted to feet of head, since the pipe elevation profile is plotted in feet. Next, starting at the first pump station A at discharge pressure P_D, a line is drawn parallel to the hydraulic gradient. The location B of the second pump station will be established at a point where the hydraulic gradient between A and B meets the vertical line at the suction pressure P_S. The process is continued to determine the location C of the third pump station.

In the above analysis we have made several simplifying assumptions. We assumed that the pressure drop per mile was constant throughout the pipeline, meaning the pipe internal diameter was uniform throughout. With variable pipe diameter or wall thickness, the hydraulic gradient slopes between pump station segments may not be the same.

6.2 Telescoping Pipe Wall Thickness

On examining the typical hydraulic gradient shown in Figure 6.2, it is evident that under steady-state operating conditions the pipe pressure decreases from pump station to the terminus in the direction of flow. Thus, the pipeline segment immediately downstream of a pump station will be subject to higher pressures such as 1000 to 1200 psi while the tail end of that segment before the next pump station (or terminus) will be subject to lower pressures in the range of 50 to 300 psi. If we use the same wall thickness throughout the pipeline, we will be underutilizing the downstream portion of the piping. Therefore, a more efficient approach would be to reduce the pipe wall thickness as we move away from a pump station toward the suction side of the next pump station or the delivery terminus.

The higher pipe wall thickness immediately adjacent to the pump station will be able to withstand the higher discharge pressure and, as the pressure reduces down the line, the lower wall thickness would be designed to withstand the lower pressures as we approach the next pump station or delivery terminus. This process of varying the wall thickness to compensate

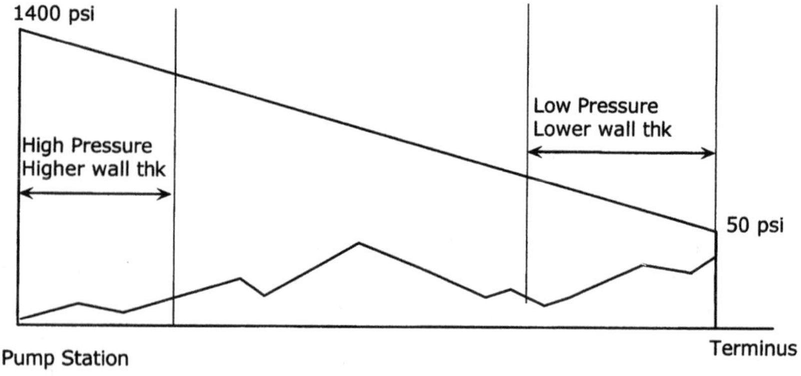

Figure 6.2 Telescoping pipe wall thickness.

for reduced pipeline pressures is referred to as telescoping pipe wall thickness.

A note of caution regarding wall thickness tapering would be appropriate here. If a pipeline has two pump stations and the second pump station is shut down for some reason, the hydraulic gradient is as shown in Figure 6.3. It can be seen that portions of the pipeline on the upstream side of the second pump station will be subject to higher pressure than when the second pump station was online. Therefore, wall thickness reductions (telescoping) implemented upstream of a pump station must be able to handle the higher pressures that result from shut-down of an intermediate pump station.

6.3 Change of Pipe Grade: Grade Tapering

In the same way that pipe wall thickness can be varied to compensate for lower pressures as we approach the next pump station or delivery terminus, the pipe grade may also be varied. Thus the high-pressure sections may be constructed of X-52 grade steel whereas the lower-pressure section may be constructed of X-42 grade pipe material, thereby reducing the total cost. This process of varying the pipe grade is referred to as grade tapering. Sometimes a combination of telescoping and grade tapering is used to minimize pipe cost. It must be noted that such wall thickness variation and pipe grade reduction to match the requirements of steady-state pressures may not always work. Consideration must be given to increased pipeline pressures when intermediate pump stations shut-down and or under upset conditions such as pump start up, valve closure, etc. These transient conditions cause surge pressures in a pipeline and therefore must be taken into account when selecting optimum wall thickness and pipe grade. This is illustrated in the modified hydraulic pressure gradient under transient conditions as depicted in Figure 6.4.

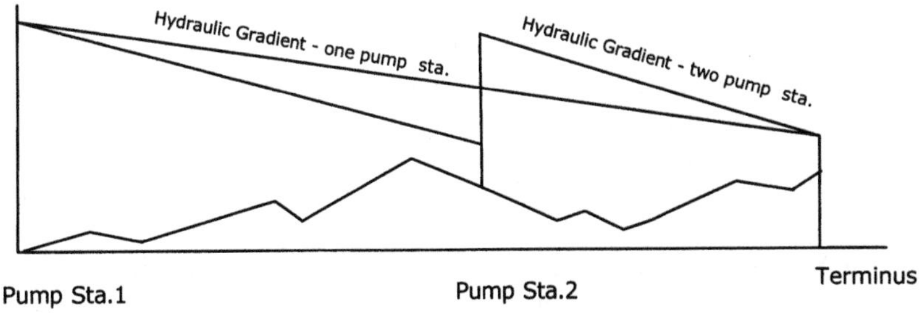

Figure 6.3 Hydraulic gradient: pump station shut-down.

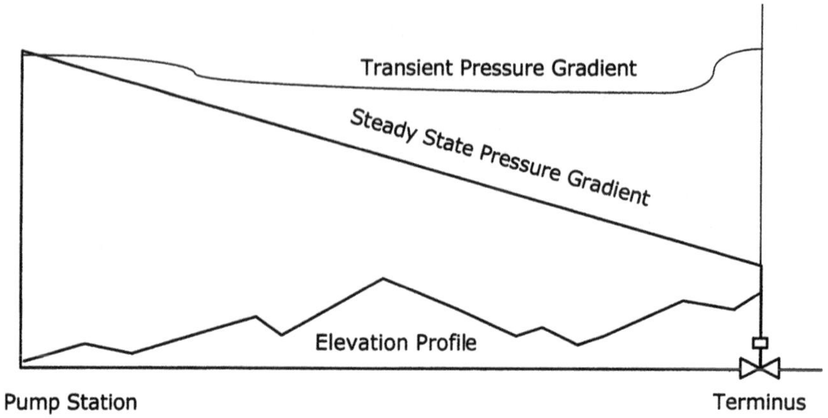

Figure 6.4 Hydraulic gradient: steady-state versus transient.

6.4 Slack Line and Open Channel Flow

Generally most pipelines flow full with no vapor space or a free liquid surface. However, under certain topographic conditions with drastic elevation changes, we may encounter pipeline sections that are partially full, called open channel flow or slack line conditions. Slack line operation may be unavoidable in some water lines, refined product and crude oil pipelines. Such a flow condition cannot be tolerated with high vapor pressure liquids and in batched pipelines. In the latter there would be intermingling of batches with disastrous consequences.

Consider a long pipeline with a very high peak at some point between the origin A and the terminus B as shown in Figure 6.5. Due to the high elevation point at C, the pressure at A must be sufficient to take care of the friction loss between A and C, and the pressure head due to elevation difference between A and C, and the minimum pressure required at the top of the hill at C to prevent vaporization of liquid. Once the liquid reaches the peak at C with the required minimum pressure, the elevation difference between C and B helps the liquid gain pressure as it flows down the hill from C to the terminus at B. The frictional pressure drop between C and B has an opposite effect to the elevation and hence the resultant pressure at the terminus B will be the difference between the elevation head and the friction head. If the elevation head between C and B is sufficiently high compared with the frictional pressure drop between C and B, the final delivery pressure at B will be higher than the minimum required at the terminus. If the delivery at B is into an atmospheric storage tank, the hydraulic gradient will be modified as shown in the dashed line for slack line conditions.

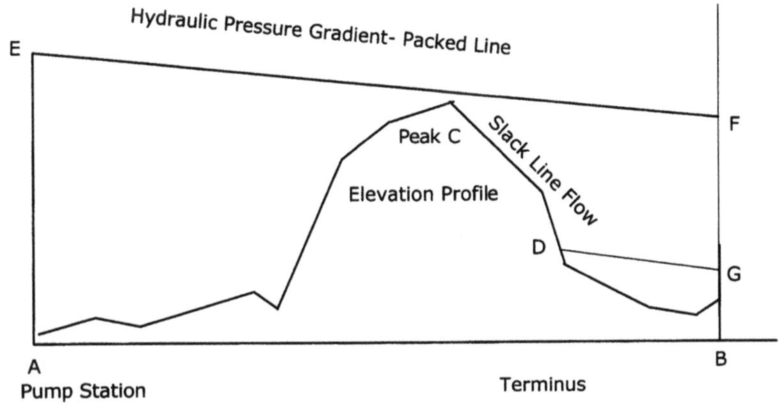

Figure 6.5 Hydraulic gradient: slack line versus packed line.

The upper hydraulic gradient depicts a packed line condition where the delivery pressure at B represented by point F is substantially higher than that required for delivery into a tank represented by point G. The lower hydraulic gradient shows that a portion of the pipeline between the peak at C and a point D will run in a partially full or slack line condition. Every point in the pipeline between C and D will be at zero gauge pressure. From D to B the pipe will run full without any slack. The slack line portion CD of the pipeline where the pipe is only partially full of the liquid is also referred to as open channel flow.

In this portion of the pipeline both liquid and vapor exists, which is an undesirable condition especially when pumping high vapor pressure liquids. Since a minimum pressure has to be maintained at the peak C, to prevent vaporization, the subsequent open channel flow in section CD of the pipeline defeats the purpose of maintaining a minimum pressure in the pipeline. In such instances, the pipeline must be operated in a packed condition (no slack line or open channel flow) by providing the necessary back pressure at B using a control valve, thus bringing the hydraulic gradient back to EF. The control valve at B should have an upstream pressure equal to the pressure that will produce the packed line hydraulic gradient showed in Figure 6.5. In crude oil and refined product pipelines where a single product is transported, slack line can be tolerated. However, if the pipeline is operated in batched mode with multiple products flowing simultaneously, slack line cannot be allowed since intermingling and consequently degradation of the different batches would occur. A batched pipeline must therefore be operated as a tight line by using a control valve to create the necessary back pressure to pack the line.

In Figure 6.5, the back pressure valve would maintain the upstream pressure corresponding to point F on the hydraulic gradient. Downstream of the valve the pressure would be lower for delivery into a storage tank.

6.5 Batching Different Liquids

Batching is the process of transporting multiple products simultaneously through a pipeline with minimal mixing. Some commingling of the batches is unavoidable at the boundary or interface between contiguous batches.

For example, gasoline, diesel, and kerosene may be shipped through a pipeline in a batched mode from a refinery to a storage terminal. Batched pipelines have to be run in turbulent mode, with velocities sufficiently high to ensure a Reynolds number over 4000. If flow were laminar ($R < 2100$), the product batches would intermingle, thereby contaminating or degrading the products. Also, as discussed earlier, batched pipelines must be run in packed conditions (no slack line or open channel flow) to avoid contamination or intermingling of batches in pipelines with drastic elevation changes such as the one shown in Figure 6.5.

In a batched pipeline, the total frictional pressure drop for a given flow rate will be calculated by adding the individual pressure drops for each product, considering its specific gravity, viscosity, and the batch length. We will illustrate this using an example.

Example Problem and Solution 6.1

Consider a 16 in. pipeline, 0.250 in. wall thickness, 100 miles long from Douglas Refinery to Hampton Terminal used to ship three products at a flow rate of 4000 bbl/hr as shown in Figure 6.6. The three batches shown represent an instantaneous snapshot condition.

First we calculate the total liquid volume in the pipeline. This is referred to as the line fill volume. The line fill volume can be calculated using the following equation:

$$\text{Line fill volume} = 5.129L(D)^2 \tag{6.5}$$

where

D = Pipe internal diameter, in.
L = Pipe length, miles

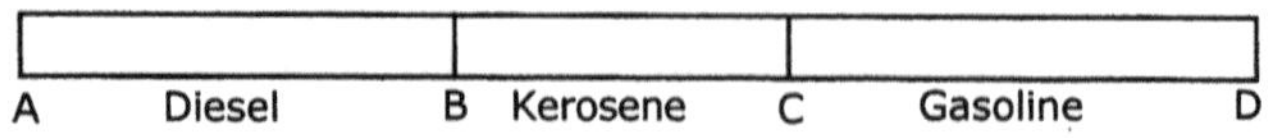

Figure 6.6 Batched pipeline.

In our example,

$$\text{Line fill} = 5.129(100)(15.5)^2 = 123{,}224 \text{ bbl}$$

Assuming the following physical properties and batch sizes for the three liquids, we can calculate the pressure drop for each liquid batch at the given flow rate:

Product	Specific gravity	Viscosity (cSt)	Batch size (bbl)
Diesel	0.85	5.9	50,000
Kerosene	0.82	2.7	30,000
Gasoline	0.74	0.7	43,224

Using the Colebrook-White equation the frictional pressure drops in the different batch segments are calculated to be as follows:

Diesel:	10.17 psi/mile	Batch length: 40.58 miles
Kerosene:	8.58 psi/mile	Batch length: 24.35 miles
Gasoline:	6.56 psi/mile	Batch length: 35.07 miles

In the above, for each liquid the pipeline length that represents the batch volume is shown. Thus the diesel batch will start at milepost 0.0 and end at milepost 40.58. Similarly the kerosene batch will start at milepost 40.58 and end at milepost $(40.58 + 24.35) = 64.93$. Finally, the gasoline batch will start at milepost 64.93 and end at milepost $(64.93 + 35.07) = 100.0$ for the snapshot configuration shown in Figure 6.6.

The batch lengths calculated above are based on 1232.24 bbl per mile of 16 in. pipe calculated using Equation 6.5. The total frictional pressure drop for the entire 100 mile pipeline is obtained by adding up the individual frictional pressure drops for each product as follows:

$$\text{Total pressure drop} = 10.17(40.58) + 8.58(24.35)$$
$$+ 6.56(35.07) = 851.68 \text{ psi.}$$

In addition to the frictional pressure drop, the elevation head and the delivery pressure are combined to calculate the total pressure required at Douglas Refinery.

In a batched pipeline, there will be some intermixing of the two adjacent products, resulting in an interface of commingled product. The volume of this interface varies with the pipe size, Reynolds number, and pipe length. Several correlations have been developed to estimate the amount of contamination in products at the interface.

As the batches arrive at the destination, density measuring instruments or densitometers (or gravitometers) monitor the density (or gravity) and flow is switched from one tank to another as appropriate. The contaminated volumes at the batch interface are diverted into a slop tank and later blended into a lower-grade product.

When batching different products such as gasoline and diesel, flow rates vary as the batches move through the pipeline, due to the changing composition of liquid in the pipeline. We will discuss this further in Chapter 7, when centrifugal pump performance is combined with system head curves.

In order to operate the batched pipeline economically, by minimizing pumping cost, there exists an optimum batch size for the various products in the pipeline system. An analysis needs to be made over a finite period, such as a week or a month, to determine the flow rates and pumping costs considering various batch sizes. The combination of batch sizes that results in the least total pumping cost, consistent with shipper and market demands, will then be the optimum batch sizes for that particular pipeline system.

6.6 Summary

We have covered hydraulic balance in pipelines with multiple pump stations and learned that hydraulic balance may not always be possible due to topographic conditions. We demonstrated an approach to determine the number of pump stations from the total pressure required, minimum pump suction pressure, and allowable pump discharge pressure based on pipeline MAOP. The advantages and cost implications of telescoping wall thickness and pipe grade tapering were discussed. Slack line and open channel flow may occur in certain cases but should be avoided in batched pipelines and when pumping high vapor pressure liquids. Also covered was the method of calculating hydraulics in a batched pipeline by analyzing a snapshot configuration of multiple products in a pipeline.

6.7 Problems

6.7.1 A pipeline 150 miles long from Beaumont pump station to a tank farm at Glendale is used to transport Alaskan North Slope crude oil (ANS crude). The pipe is 20 in. in outside diameter and constructed of X-52 steel. It is desired to operate the system at ANSI 600 pressure level (1440 psi). The pipeline profile is such that there are two peaks located between Beaumont and Glendale. The first peak occurs at milepost 65.0 at an elevation of 1500 ft. The second peak is located at milepost 110.0 at an elevation of 2500 ft. Beaumont has an elevation of 350 ft and Glendale is situated at an elevation of 650 ft. During the initial phase of operation, 6000 bbl/hr of ANS crude will be pumped at a temperature of 60°F and delivered to Glendale tankage at a pressure of 30 psi. The specific gravity and viscosity of ANS crude at 60°F may be assumed to be 0.895 and 43 cSt respectively.

(a) Determine the minimum wall thickness required to operate the pipeline system at a pressure of 1400 psi.
(b) At a flow rate of 6000 bbl/hr, how many pump stations would be required?
(c) During the second phase it is planned to expand the capacity of pipeline to 9000 bbl/hr. How many additional pump stations would be required?
(d) Assuming 80% pump efficiency, calculate the total pumping HP required during the initial phase and under the expansion scenario. Use a minimum suction pressure of 50 psi at each pump station.

6.7.2 The pipeline described in Problem 6.7.1 is used to batch ANS crude along with a light crude (0.85 specific gravity and 15 cSt viscosity at 60°F). Determine the optimum batch sizes to reduce pumping costs based on a 30 day operation. Consider a flow rate of 6000 bbl/hr.

6.7.3 A batched pipeline, 60 km long, is used to ship three grades of refined products. An instantaneous configuration shows an 8000 m^3 batch of gasoline followed by 10,000 m^3 of diesel, the remainder of the line consisting of a batch of kerosene. The pipe size is 500 mm diameter and 10 mm wall thickness and it traverses essentially flat terrain. Assume 8 MPa for the operating pressure and a pipeline delivery pressure of 300 kPa. Calculate the total pressure drop in the pipeline at a flow rate of 1800 m^3/hr. The specific gravity and viscosities of the three products at the flowing temperature of 20°C are:

	Specific gravity	Viscosity (cSt)
Gasoline	0.74	0.65
Kerosene	0.82	1.5
Diesel	0.85	5.9

7

Pump Analysis

In this chapter we will discuss centrifugal pumps used in liquid pipelines. Although other types of pumps such as rotary pumps and piston pumps are sometimes used, the majority of pipelines today are operated with single- and multi-stage centrifugal pumps. We will cover the basic design of centrifugal pumps, their performance characteristics and how the performance may be modified by changing pump impeller speeds and trimming impellers. We introduce Affinity Laws for centrifugal pumps, the importance of net positive suction head (NPSH) and how to calculate horsepower requirements when pumping different liquids. Also covered is the pump performance correction for high-viscosity liquids using the Hydraulic Institute chart. We will also illustrate how the performance of two or more pumps in series or parallel configuration can be determined and how to estimate the operating point for a pipeline by using the system head curve.

7.1 Centrifugal Pumps Versus Reciprocating Pumps

Pumps are needed to raise the pressure of a liquid in a pipeline so that the liquid may flow from the beginning of the pipeline to the delivery terminus at the required flow rate and pressure. As the flow rate increases more pump pressure will be required.

Over the years pipelines have been operated with centrifugal as well as reciprocating or positive displacement (PD) pumps. In this chapter we will concentrate mainly on centrifugal pumps, as these are used extensively in most liquid pipelines that transport water, petroleum products, and chemicals. PD pumps are discussed as well, since they are used for liquid injection lines in oil pipeline gathering systems.

Centrifugal pumps develop and convert the high liquid velocity into pressure head in a diffusing flow passage. They generally have a lower efficiency than PD pumps such as reciprocating and gear pumps. However, centrifugal pumps can operate at higher speeds to generate higher flow rates and pressures. Centrifugal pumps also have lower maintenance requirements than PD pumps.

Positive displacement pumps such as reciprocating pumps operate by forcing a fixed volume of liquid from the inlet to the outlet of the pump. These pumps operate at lower speeds than centrifugal pumps. Reciprocating pumps cause intermittent flow. Rotary screw pumps and gear pumps are also PD pumps, but operate continuously unlike reciprocating pumps. Positive displacement pumps are generally larger in size and more efficient compared with centrifugal pumps, but require higher maintenance.

Modern pipelines are mostly designed with centrifugal pumps in preference to PD pumps. This is because there is more flexibility in volumes and pressures with centrifugal pumps. In petroleum pipeline installations where liquid from a field gathering system is injected into a main pipeline, PD pumps may be used.

The performance of a centrifugal pump is depicted as a curve that shows variation in pressure head at different flow rates. The pump head characteristic curve shows the pump head developed on the vertical axis, while the flow rate is shown on the horizontal axis. This curve may be referred to as the H-Q curve or the head-capacity curve. Pump companies use the term capacity when referring to flow rate. Throughout this section we use capacity and flow rate interchangeably when dealing with pumps.

Generally, pump performance curves are plotted for water as the liquid pumped. The head is measured in feet of water and flow rate is shown in gal/min. In addition to the head versus flow rate curve, two other curves are commonly shown on a typical pump performance chart. These are pump efficiency versus flow rate and pump brake horsepower (BHP) versus flow rate. Figure 7.1 shows typical centrifugal pump performance curves consisting of head, efficiency and BHP plotted against flow rate (or capacity). You will also notice another curve referred to as NPSH plotted against flow rate. NPSH will be discussed later in this chapter.

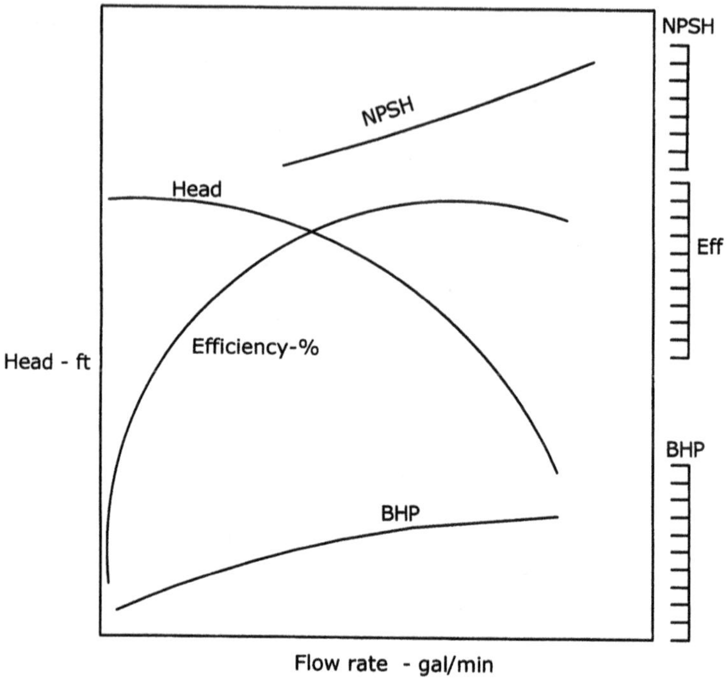

Figure 7.1 Centrifugal pump performance.

In summary, typical centrifugal pump characteristic curves include the following four curves:

Head versus flow rate
Efficiency versus flow rate
BHP versus flow rate
NPSH versus flow rate

The above performance curves for a particular model pump are generally plotted for a particular pump impeller size and speed (example: 10 in. impeller, 3560 RPM).

In addition to the above four characteristic curves, pump head curves drawn for different impeller diameters and iso-efficiency curves are sometimes encountered. When pumps are driven by variable-speed electric motors or engines, the head curves may also be shown at various pump speeds. Pump performance at various impeller sizes and speeds will be discussed in detail later in this chapter.

A PD pump continuously pumps a fixed volume at various pressures. It is able to provide any pressure required at a fixed flow rate. This flow rate

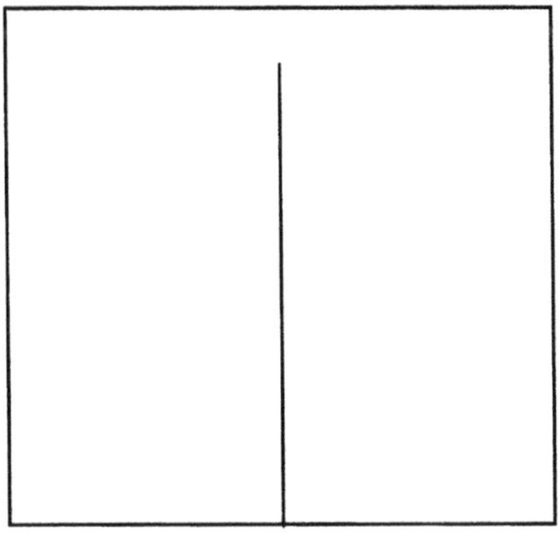

Figure 7.2 Positive displacement pump performance.

depends on the geometry of the pump such as bore and stroke. A typical PD pump pressure-volume curve is shown in Figure 7.2.

7.2 Pump Head Versus Flow Rate

For a centrifugal pump, the head-capacity (flow rate) variation is shown in Figure 7.3.

The head versus flow rate curve (H-Q curve) is generally referred to as a drooping head curve that starts at the highest value (shut-off head) at zero flow rate. The head decreases as the flow rate through the pump increases. The trailing point of the curve represents the maximum flow and the corresponding head that the pump can generate.

Pump head is always plotted in feet of head of water. Therefore the pump is said to develop the same head in feet of liquid, regardless of the liquid. If a particular pump develops 2000 ft head at 3000 gal/min flow rate, we can calculate the pressure developed in psi when pumping water versus another liquid such as gasoline:

$$\text{With water, pressure developed} = 2000 \times 1/2.31 = 866 \text{ psi}$$

$$\text{With gasoline, pressure developed} = 2000 \times 0.74/2.31 = 641 \text{ psi}$$

This head versus flow rate relationship holds good for this particular pump that has a fixed impeller diameter D and operates at a fixed speed N. The diameter of the impeller generally can be increased within a certain range of sizes, depending on the internal cavity that houses the pump impeller. Thus,

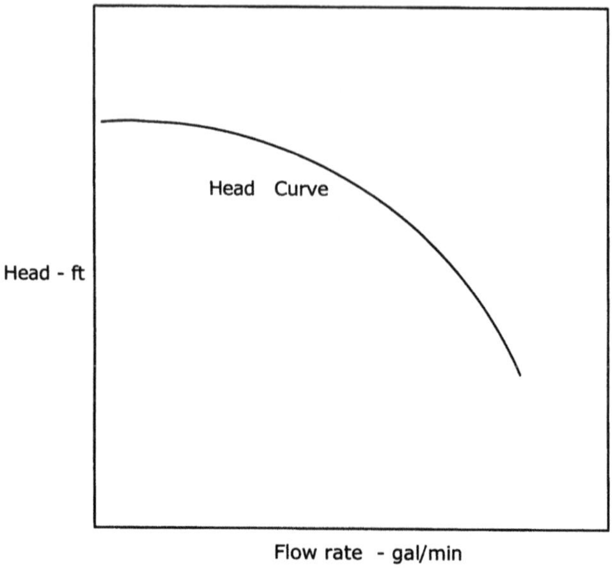

Figure 7.3 Centrifugal pump performance head versus flow rate.

if the H-Q curve in Figure 7.3 is based on a 10 in. impeller diameter, similar curves can be generated for 9 in. impeller and 12 in. impellers representing the minimum and maximum sizes available for this particular pump. The minimum and maximum impeller sizes are defined by the pump vendor for the particular pump model. The 9 in. curve and 12 in. curves would be parallel to the 10 in. curve as shown in Figure 7.4. The variations in pump H-Q curves with impeller diameter follow the Affinity Laws, discussed in a later section of this chapter.

Similar to H-Q variations with pump impeller diameter, curves can be generated for varying pump impeller speeds. If the impeller diameter is kept constant at 10 in. and the initial H-Q curve was based on a pump speed of 3560 RPM (typical induction motor speed), by varying the pump speed we can generate a family of parallel curve as shown in Figure 7.5.

It will be observed that the H-Q variations with pump impeller diameter and pump speed are similar. This is due to the Affinity Laws that the pump performance is based on. We discuss Affinity Laws and prediction of pump performance at different diameters and speeds later in this chapter.

A pump may develop the head in stages. Thus single-stage and multi-stage pumps are used depending upon the head required. A single-stage

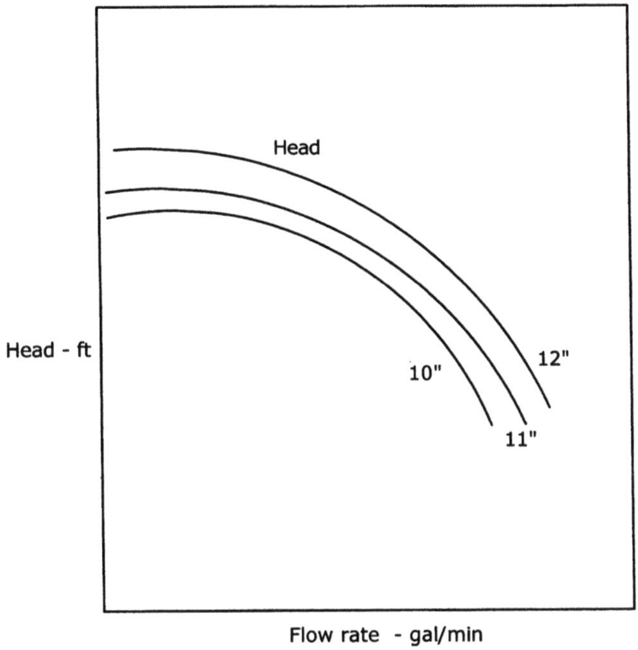

Figure 7.4 Head versus flow rate for different impeller sizes.

pump may generate 200 ft head at 3000 gal/min. A three-stage pump of this design will generate 600 ft head at same flow rate and pump speed. An application that requires 2400 ft head at 3000 gal/min may be served by a multi-stage pump with each stage providing 400 ft of head. De-staging is the process of reducing the active number of stages in a pump to reduce the total head developed. Thus, the six-stage pump discussed above may be de-staged to four stages if we need only 1600 ft head at 3000 gal/min.

7.3 Pump Efficiency Versus Flow Rate

The variation of the efficiency E of a centrifugal pump with flow rate Q is as shown in Figure 7.6. It can be seen that the efficiency starts off at zero value under shut-off conditions (zero flow rate) and rises to a maximum value at some flow rate. After this point, with increase in flow rate the efficiency drops off to some value lower than the peak efficiency. Generally centrifugal pump peak efficiencies are approximately 80–85%. It must be remembered in all these discussions that we are referring to a centrifugal

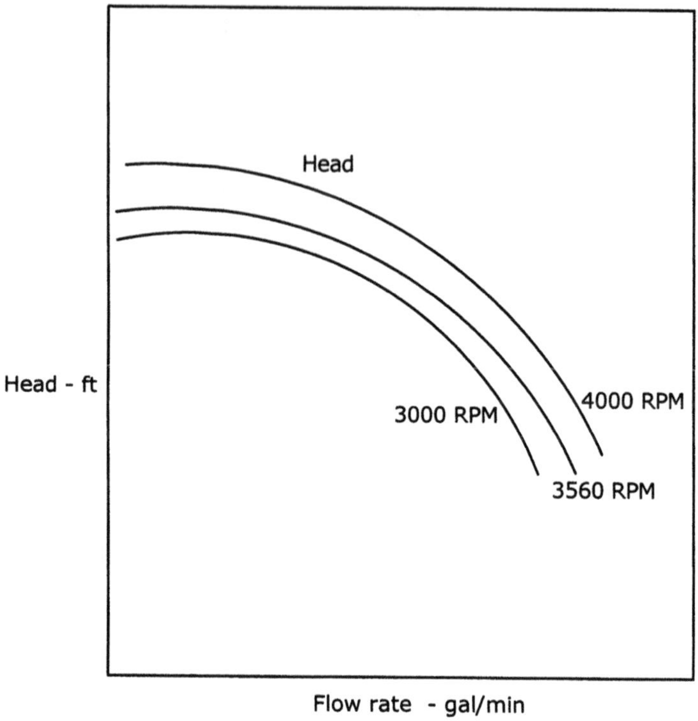

Figure 7.5 Head versus flow rate at different speeds.

pump performance with water as the liquid being pumped. When heavy, viscous liquids are pumped, these water-based efficiencies will be reduced by a correction factor. Viscosity-corrected pump performance using the Hydraulic Institute charts is discussed in Section 7.8. The flow rate at which the maximum efficiency occurs in a pump is referred to as the best efficiency point (BEP). This flow rate and the corresponding head from the H-Q curve is thus the best operating point on the pump curve since the highest pumping efficiency is realized at this flow rate. In the combined H-Q and efficiency versus flow rate (E-Q) curves shown in Figure 7.7, the BEP is shown as a small triangle.

When choosing a centrifugal pump for a particular application we try to get the operating point as close as possible to the BEP. In order to allow for future increase in flow rates, the initial operating point is chosen slightly to the left of the BEP. This will ensure that with increase in pipeline throughput the operating point on the pump curve will move to the right, which would result in a slightly better efficiency.

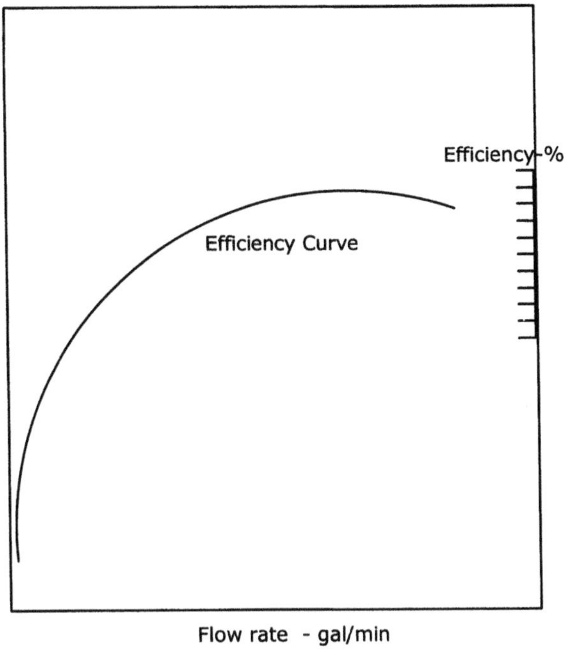

Figure 7.6 Centrifugal pump performance: efficiency versus flow rate.

7.4 BHP Versus Flow Rate

From the H-Q and E-Q curves we can calculate the BHP required at every point along the curve using Equation (5.17) as follows:

$$\text{Pump BHP} = \frac{Q\,H\,Sg}{3960(E)} \tag{7.1}$$

where

Q = Pump flow rate, gal/min
H = Pump head, ft
E = Pump efficiency, as a decimal value less than 1.0
Sg = Liquid specific gravity (for water Sg = 1.0)

In SI units, power in kW can be calculated as follows:

$$\text{Power kW} = \frac{Q\,H\,Sg}{367.46\,(E)} \tag{7.2}$$

where

Q = Pump flow rate, m^3/hr
H = Pump head, m

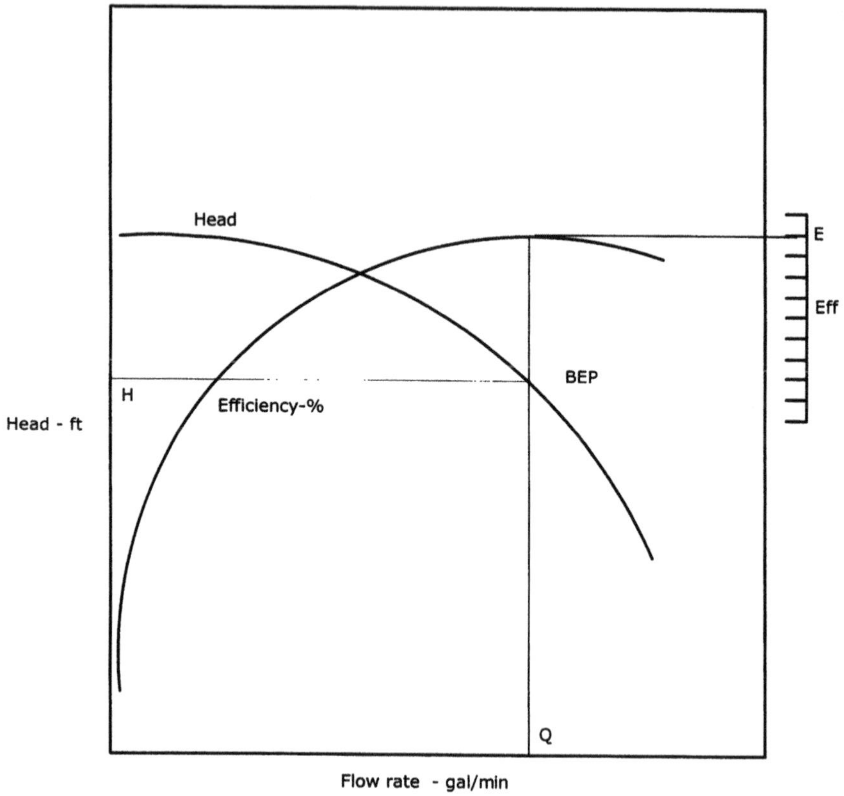

Figure 7.7 Best efficiency point.

E = Pump efficiency, as a decimal value less than 1.0
Sg = Liquid specific gravity (for water Sg = 1.0)

For example, if the BEP on a particular pump curve occurs at a flow rate of 3000 gal/min, 2500 ft head at 85% efficiency, the BHP at this flow rate for water is calculated from Equation (7.1) as

$$\text{Pump BHP} = 3000 \times 2500 \times 1.0/(3960 \times 0.85) = 2228$$

This represents the BHP with water at 3000 gal/min flow rate. Similarly, BHP can be calculated at various flow rates from zero to 4000 gal/min (assumed maximum pump flow) by reading the corresponding head and efficiency values from the H-Q curve and E-Q curve and using Equation (7.1) as above. The BHP versus flow rate can then be plotted as shown in Figure 7.8. The BHP versus flow rate curve can also be shown on the same plot as H-Q and E-Q curves as shown in Figure 7.1.

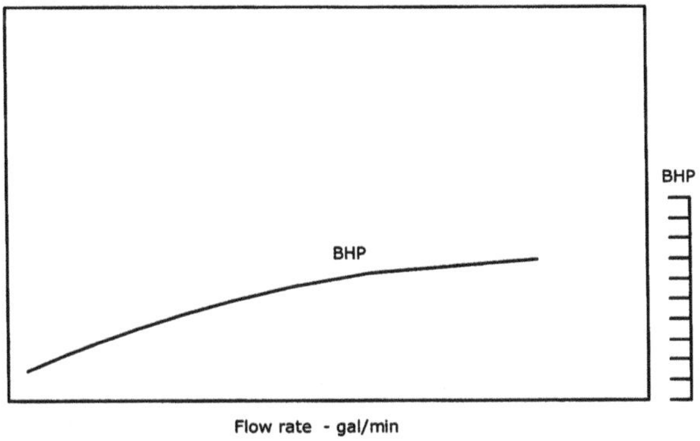

Figure 7.8 BHP versus flow rate.

As discussed in Section 5.6.2, the BHP calculated above is the brake horsepower demanded by the pump. An electric motor driving the pump will have an efficiency that ranges from 95% to 98%. Therefore the electric motor HP required is the pump BHP divided by the motor efficiency as follows

$$\begin{aligned}\text{Motor HP} &= \text{Pump BHP/Motor efficiency}\\ &= 2228/0.96\\ &= 2321 \text{ HP, based on 96\% motor efficiency}\end{aligned}$$

In determining the motor size required, we must first calculate the maximum HP demanded by the pump, which will generally (but not necessarily) be at the highest flow rate. The BHP at this flow rate will then be divided by the electric motor efficiency to get the maximum motor HP required. The next available standard size motor will be selected for the application.

Suppose that for the above pump 4000 gal/min is the maximum flow rate through the pump and the head and efficiency are 1800 ft and 76%, respectively. We will calculate the maximum BHP required as

$$\text{Maximum pump BHP} = \frac{4000 \times 1800 \times 1.0}{3960 \times 0.76} = 2392.34$$

$$\text{Motor HP} = \frac{2392.34}{0.96} = 2492$$

Therefore a 2500 HP electric motor will be adequate for this case. Before we leave the subject of pump BHP and electric motor HP, we must mention that electric motors have built into their nameplate rating a service factor. The service factor for most induction motors is in the range of 1.10 to 1.15. This simply means that if the motor's rated HP is 2000, the 1.15 service factor allows the motor to supply up to a maximum of 1.15×2000 or 2300 HP without burning up the motor windings. However, it is not advisable to use this extra 300 HP on a continuous basis, as explained below.

At a 1.15 service factor rating, the motor windings will be taking on an additional 15% electric current compared with their rated current (amps). This translates to extra heat that will shorten the life of the windings. Even though the extra current is only 15%, the heat generated will be over 32%, since electrical heating is proportional to the square of the current ($1.15^2 = 1.3225$). However, while continuous operation at the service factor rating is discouraged, in an emergency this additional 15% motor HP above the rated HP may be used.

7.5 NPSH Versus Flow Rate

In addition to the three pump curves discussed previously—for head versus capacity (H-Q), efficiency versus capacity (E-Q), and BHP versus capacity (BHP-Q)—the pump performance data will include a fourth curve for net positive suction head (NPSH) versus capacity. This curve is generally located above the head, efficiency, and BHP curves as shown in Figure 7.1.

The NPSH curve shows the variation in the minimum net positive suction head at the impeller suction versus the flow rate. The NPSH increases at a faster rate as the flow rate increases. NPSH is defined as the net positive suction head required at the pump impeller suction to prevent pump cavitation at any flow rate. It represents the resultant positive pressure at pump suction after accounting for frictional loss and liquid vapor pressure. NPSH is discussed in detail in Section 7.12. At present, we can say that as pump flow rate increases the NPSH required also increases. We will perform calculations to compare NPSH required (per pump curve) versus actual NPSH available based on pump suction piping and other parameters that affect the available suction pressure at the pump.

7.6 Specific Speed

The specific speed of a centrifugal pump is a parameter based on the impeller speed, flow rate, and head at the best efficiency. It is used for comparing geometrically similar pumps and for classifying the different types of centrifugal pumps.

Specific speed may be defined as the speed at which a geometrically similar pump must be run such that it produces a head of 1 ft at a flow rate of 1 gal/min. Mathematically, specific speed is defined as

$$N_S = NQ^{1/2}/H^{3/4} \tag{7.3}$$

where

N_S = Pump specific speed
N = Pump impeller speed, RPM
Q = Flow rate or capacity, gal/min
H = Head, ft

Both Q and H are measured at the BEP for the maximum impeller diameter. The head H is measured per stage for a multi-stage pump. Low specific speed is associated with high-head pumps, while high specific speed is found with low-head pumps.

Another related term, the suction specific speed, is defined as follows:

$$N_{SS} = NQ^{1/2}/(NPSH_R)^{3/4} \tag{7.4}$$

where

N_{SS} = Suction specific speed
N = Pump impeller speed, RPM
Q = Flow rate or capacity, gal/min
$NPSH_R$ = NPSH required at the BEP

When applying the above equations to calculate the pump specific speed and suction specific speed, use the full Q value for single or double suction pumps for N_S calculation. For N_{SS} calculation, use one-half the Q value for double suction pumps.

Example Problem 7.1

Calculate the specific speed of a five-stage double suction centrifugal pump, 12 in. diameter impeller, that when operated at 3560 RPM generates a head of 2200 ft at a capacity of 3000 gal/min at the BEP on the head capacity curve. If the NPSH required is 25 ft, calculate the suction specific speed.

Solution

$$N_S = N\,Q^{1/2}/H^{3/4}$$

$$= 3560(3000)^{1/2}/(2200/5)^{3/4} = 2030$$

The suction specific speed is

$$N_{SS} = N\,Q^{1/2}/NPSH_R^{3/4}$$

$$= 3560(3000/2)^{1/2}/(25)^{3/4} = 12{,}332$$

Table 7.1 Specific Speeds of Centrifugal Pumps

Description	Application	Specific speed, Ns
Radial vane	Low capacity, high head	500–1000
Francis-screw type	Medium capacity, medium head	1000–4000
Mixed-flow type	Medium to high capacity, low to medium head	4000–7000
Axial-flow type	High capacity, low head	7000–20,000

Centrifugal pumps are generally classified as

(a) Radial flow
(b) Axial flow
(c) Mixed flow

Radial flow pumps develop head by centrifugal force. Axial flow pumps on the other hand develop the head due to the propelling or lifting action of the impeller vanes on the liquid. Radial flow pumps are used when high heads are required, while axial flow and mixed flow pumps are mainly used for low-head/high-capacity applications. Table 7.1 lists the specific speed range for centrifugal pumps.

7.7 Affinity Laws: Variation with Impeller Speed and Diameter

Each family of pump performance curves for head, efficiency, and BHP versus capacity is specific to a particular size of impeller and pump model. Thus, Figure 7.1 depicts pump performance data for a 10 in. diameter pump impeller. This particular pump may be able to accommodate a larger impeller, such as 12 in. in diameter. Alternatively, a smaller, 9 in. impeller may also be fitted in this pump casing. The range of impeller sizes is dependent on the pump case design and will be defined by the pump vendor. Since centrifugal pumps generate pressure due to centrifugal force, it is clear that a smaller impeller diameter will generate less pressure at a given flow rate than a larger impeller at the same speed. If the performance data for a 10 in. impeller is available, we can predict the pump performance when using a larger or smaller size impeller by means of the centrifugal pump Affinity Laws.

The same pump with a 10 in. impeller may be operated at a higher or lower speed to provide higher or lower pressure. Most centrifugal pumps are driven by constant-speed electric motors. A typical induction motor in the United States operates at 60 Hz and will have a synchronous speed of

3600 RPM. With some slip, the induction motor would probably run at 3560 RPM. Thus, most constant-speed motor-driven pumps have performance curves based on 3560 RPM. Some slower-speed pumps operate at 1700 to 1800 RPM. If a variable-speed motor is used, the pump speed can be varied from, say, 3000 to 4000 RPM. This variation in speed produces variable head versus capacity values for the same 10 in. impeller. Given the performance data for a 10 in. impeller at 3560 RPM, we can predict the pump performance at different speeds ranging from 3000 to 4000 RPM using the centrifugal pump Affinity Laws as described next.

The Affinity Laws for centrifugal pumps are used to predict pump performance for changes in impeller diameter and impeller speed. According to the Affinity Laws, for small changes in impeller diameter the flow rate (pump capacity) Q is directly proportional to the impeller diameter. The pump head H, on the other hand, is directly proportional to the square of the impeller diameter. Since the BHP is proportional to the product of flow rate and head (see Equation 7.1), BHP will vary as the third power of the impeller diameter.

The Affinity Laws are represented mathematically as follows:
For impeller diameter change:

$$Q_2/Q_1 = D_2/D_1 \tag{7.5}$$

$$H_2/H_1 = (D_2/D_1)^2 \tag{7.6}$$

where

Q_1, Q_2 = Initial and final flow rates
H_1, H_2 = Initial and final heads
D_1, D_2 = Initial and final impeller diameters

From the H-Q curve corresponding to the impeller diameter D_1 we can select a set of H-Q values that cover the entire range of capabilities from zero to the maximum possible. Each value of Q and H on the corresponding H-Q curve for the impeller diameter D_2 can then be computed using the Affinity Law ratios per Equations (7.5) and (7.6).

Similarly, Affinity Laws state that for the same impeller diameter, if pump speed is changed, flow rate is directly proportional to the speed, while the head is directly proportional to the square of the speed. As with diameter change, the BHP is proportional to the third power of the impeller speed. This is represented mathematically as follows:
For impeller speed change:

$$Q_2/Q_1 = N_2/N_1 \tag{7.7}$$

$$H_2/H_1 = (N_2/N_1)^2 \tag{7.8}$$

where

Q_1, Q_2 = Initial and final flow rates
H_1, H_2 = Initial and final heads
N_1, N_2 = Initial and final impeller speeds

Note that the Affinity Laws for speed change are exact. However, the Affinity Laws for impeller diameter change are only approximate and valid for small changes in impeller sizes. The pump vendor must be consulted to verify that the predicted values using Affinity Laws for impeller size changes are accurate or whether any correction factors are needed. With speed and impeller size changes, the efficiency versus flow rate can be assumed to be the same.

An example using the Affinity Laws will illustrate how the pump performance can be predicted for impeller diameter change as well as for impeller speed change.

Example Problem 7.2

The head and efficiency versus capacity data for a centrifugal pump with a 10 in. impeller is as shown below.

Q, gal/min	0	800	1600	2400	3000
H, ft	3185	3100	2900	2350	1800
E, %	0.0	55.7	78.0	79.3	72.0

The pump is driven by a constant-speed electric motor at a speed of 3560 RPM.

(a) Determine the performance of this pump with an 11 in. impeller, using Affinity Laws.
(b) If the pump drive were changed to a variable frequency drive (VFD) motor with a speed range of 3000 to 4000 RPM, calculate the new H-Q curve for the maximum speed of 4000 RPM with the original 10 in. impeller.

Solution

(a) Using Affinity Laws for impeller diameter changes, the multiplying factor for flow rate is factor $= 11/10 = 1.1$ and the multiplier for head is $(1.1)^2 = 1.21$. Therefore, we will generate a new set of Q and H values for the 11 in. impeller by multiplying the given Q values by the factor 1.1 and the H values by the factor 1.21 as follows:

Q, gal/min	0	880	1760	2640	3300
H, ft	3854	3751	3509	2844	2178

The above flow rate and head values represent the predicted performance of the 11 in. impeller. The efficiency versus flow rate curve for the 11 in. impeller will be approximately the same as that of the 10 in. impeller.

(b) Using Affinity Laws for speeds, the multiplying factor for the flow rate is factor = 4000/3560 = 1.1236 and the multiplier for head is $(1.1236)^2 = 1.2625$. Therefore we will generate a new set of Q and H values for the pump at 4000 RPM by multiplying the given Q values by the factor 1.1236 and the H values by the factor 1.2625 as follows:

Q, gal/min	0	899	1798	2697	3371
H, ft	4021	3914	3661	2967	2273

The above flow rates and head values represent the predicted performance of the 10 in. impeller at 4000 RPM. The new efficiency versus flow rate curve will be approximately the same as the given curve for 3560 RPM.

Thus, using Affinity Laws, we can determine whether we should increase or decrease the impeller diameter to match the requirements of flow rate and head for a specific pipeline application.

For example, suppose hydraulic calculations for a particular pipeline indicate we need a pump that can provide a head of 2000 ft at a flow rate of 2400 gal/min. The data for the pump in Example Problem 7.2 shows that a 10 in. impeller produces 2350 ft head at a flow rate of 2400 gal/min. If this is a constant-speed pump, it is clear that in order to get 2000 ft head, the current 10 in. impeller needs to be trimmed in diameter to reduce the head from 2350 ft to the 2000 ft required. The Example Problem 7.4 below will illustrate how the amount of impeller trim is calculated.

7.8 Effect of Specific Gravity and Viscosity on Pump Performance

The performance of a pump as reported by the pump vendor is always based on water as the pumped liquid. Hence the head versus capacity curve, the efficiency versus capacity curve, BHP versus capacity curve, and NPSH required versus capacity curve are all applicable only when the liquid pumped is water (specific gravity of 1.0) at standard conditions, usually 60°F. When pumping a liquid such as crude oil with specific gravity of 0.85 or a refined product such as gasoline with a specific gravity of 0.74, both the H-Q and E-Q curves may still apply for these products if the viscosities are sufficiently low (less than 4.3 cSt for small pumps up to 100 gal/min capacity and less than 40 SSU for larger pumps up to 10,000 gal/min) according to Hydraulic Institute standards. Since BHP is a function of the specific gravity, from Equation (7.1) it is clear that the BHP versus flow rate will be different for other liquids compared with water. If the liquid pumped has a higher viscosity, in the range of 4.3 to 3300 cSt, the head, efficiency, and BHP curves based on water must all be corrected for the high viscosity. The Hydraulic Institute has published viscosity correction charts that can be applied to correct the water performance curves to produce viscosity-corrected curves. See Figure 7.9 for details of this chart. For any application involving high-viscosity liquids, the pump vendor should be given the liquid properties. The viscosity-corrected performance curves will be supplied by the vendor as part of the pump proposal. As an end user you may also use these charts to generate the viscosity-corrected pump curves.

The Hydraulic Institute method of viscosity correction requires determining the BEP values for Q, H, and E from the water performance curve. This is called the 100% BEP point. Three additional sets of Q, H, and E values are obtained at 60%, 80%, and 120% of the BEP flow rate from the water performance curve. From these four sets of data, the Hydraulic Institute chart can be used to obtain the correction factors C_q, C_h, and C_e for flow, head, and efficiency for each set of data. These factors are used to multiply the Q, H, and E values from the water curve, thus generating corrected values of Q, H, and E for 60%, 80%, 100%, and 120% BEP values. Example Problem 7.3 illustrates the Hydraulic Institute method of viscosity correction. Note that for multi-stage pumps, the values of H must be per stage.

Example Problem 7.3

The water performance of a single-stage centrifugal pump for 60%, 80%, 100%, and 120% of the BEP is as shown below:

Q, gal/min	450	600	750	900
H, ft	114	108	100	86
E, %	72.5	80.0	82.0	79.5

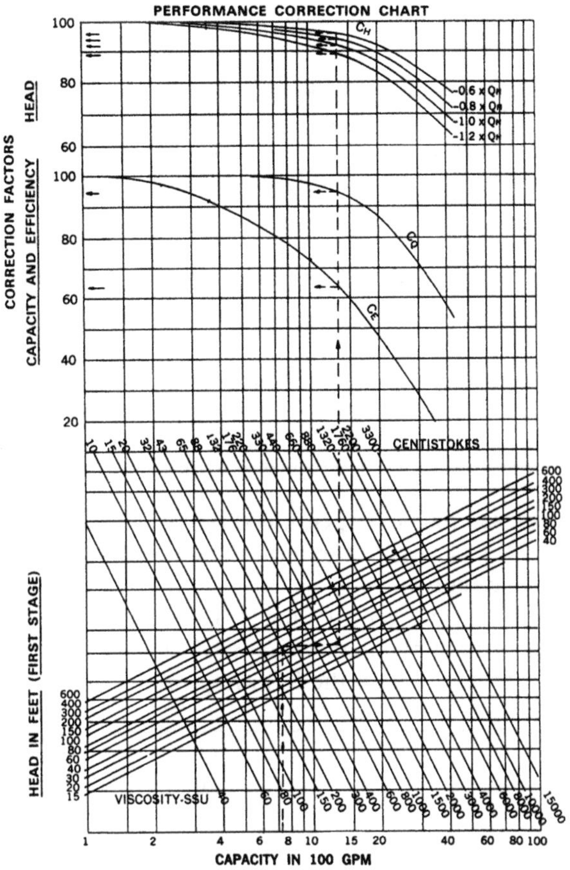

Figure 7.9 Viscosity correction chart. (Courtesy of Hydraulic Institute, Parsippany, NJ; www.pumps.org.)

Calculate the viscosity-corrected pump performance when pumping oil with a specific gravity of 0.90 and a viscosity of 1000 SSU at pumping temperature.

Solution

By inspection, the BEP for this pump curve is

$Q = 750$
$H = 100$
$E = 82$

We first establish the four sets of capacities to correspond to 60%, 80%, 100%, and 120%. These have already been given as 450, 600, 750, and

900 gal/min. Since the head values are per stage, we can directly use the BEP value of head along with the corresponding capacity to enter the Hydraulic Institute viscosity correction chart at 750 gal/min on the lower horizontal scale, go vertically from 750 gal/min to the intersection point on the line representing the 100 ft head curve and then horizontally to intersect the 1000 SSU viscosity line, and finally vertically up to intersect the three correction factor curves C_e, C_q, and C_h.

From the Hydraulic Institute chart of correction factors (Figure 7.9) we obtain the values of C_q, C_h, and C_e for flow rate, head, and efficiency as follows:

C_q	0.95	0.95	0.95	0.95
C_h	0.96	0.94	0.92	0.89
C_e	0.635	0.635	0.635	0.635

corresponding to Q values of 450 (60% of Q_{NW}), 600 (80% of Q_{NW}), 750 (100% of Q_{NW}), and 900 (120% of Q_{NW}). The term Q_{NW} is the BEP flow rate from the water performance curve.

Using these correction factors, we generate the Q, H, and E values for the viscosity-corrected curves by multiplying the water performance value of Q by C_q, H by C_h, and E by C_e and obtain the following result:

Q_V	427	570	712	855
H_V	109.5	101.5	92.0	76.5
E_V	46.0	50.8	52.1	50.5
BHP_V	23.1	25.9	28.6	29.4

The last row of values for viscous BHP was calculated using Equation (7.1) as follows:

$$BHP_V = (Q_V)(H_V)0.9/(39.60 \times E_V) \quad (7.9)$$

where Q_V, H_V, and E_V are the viscosity-corrected values of capacity, head, and efficiency tabulated above.

Note that when using the Hydraulic Institute chart for obtaining the correction factors, two separate charts are available. One chart applies to small pumps up to 100 gal/min capacity and head per stage of 6 to 400 ft. The other chart applies to larger pumps with capacity between 100 and 10,000 gal/min and head range of 15 to 600 ft per stage. Also, remember that when data is taken from a water performance curve, the head has to be corrected per stage, since the Hydraulic Institute charts are based on head in ft per stage rather than the total pump head. Therefore, if a six-stage pump has a BEP at 2500 gal/min and 3000 ft of head with an efficiency of 85%, the head per stage to be used with the chart will be 3000 divided by $6 = 500$ ft. The total head (not per stage) from the water curve can then be multiplied by the correction factors from the Hydraulic Institute charts to obtain the viscosity-corrected head for the six-stage pump.

7.9 Pump Curve Analysis

So far we have discussed the performance of a single pump. When transporting liquid through a pipeline we may need to use more than one pump at a pump station to provide the necessary flow rate or head requirement. Pumps may be operated in series or parallel configurations. Series pumps are generally used for higher heads and parallel pumps for increased flow rates.

For example, let us assume that pressure drop calculations indicate that the originating pump station on a pipeline requires 900 psi differential pressure to pump 3000 gal/min of gasoline with specific gravity 0.736. Converting to customary pump units we can state that we require a pump that can provide the following:

$$\text{Head} = \frac{900 \times 2.31}{0.736} = 2825 \text{ ft}$$

at 3000 gal/min flow rate. We have two options here. We could select from a pump vendors catalog one large pump that can provide 2825 ft head at 3000 gal/min or select two smaller pumps that can provide 1413 ft at 3000 gal/min. We would use these two smaller pumps in series to provide the required head, since for pumps operated in series the same flow rate goes through each pump and the resultant head is the sum of the heads generated by each pump. Alternatively, we could select two pumps that can provide 2825 ft at 1500 gal/min each. In this case these pumps would be operated in parallel. In parallel operation, the flow rate is split between the pumps, while each pump produces the same head. Series and parallel pump configuration are illustrated in Figure 7.10.

The choice of series or parallel pumps for a particular application depends on many factors, including pipeline elevation profile, as well as operational flexibility. Figure 7.11 shows the combined performance of two identical pumps in series, versus parallel configuration. It can be seen that parallel pumps are used when we need larger flows. Series pumps are used when we need higher heads than each individual pump. If the pipeline elevation profile is essentially flat, the pump pressure is required mainly to overcome the pipeline friction. On the other hand, if the pipeline has drastic elevation changes, the pump head generated is mainly for the static

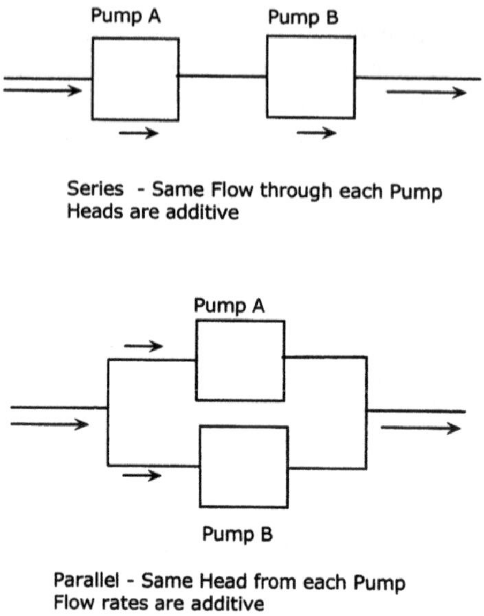

Figure 7.10 Pumps in series and parallel.

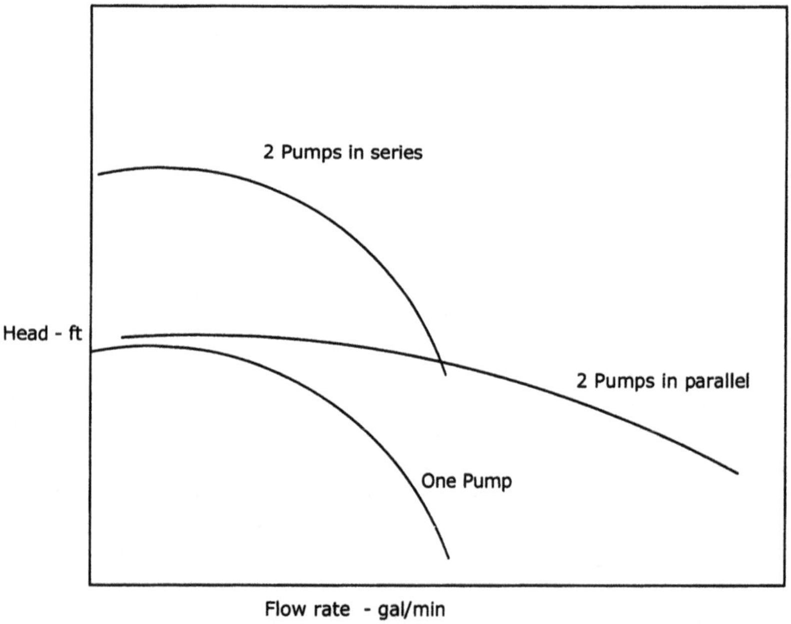

Figure 7.11 Pump performance: series and parallel.

lift and to a lesser extent for pipe friction. In the latter case, if two pumps are used in series and one shuts down, the remaining pump alone will only be able to provide half the head and therefore will not be able to provide the necessary head for the static lift at any flow rate. If the pumps were configured in parallel, then shutting down one pump will still allow the other pump to provide the necessary head for the static lift at half the previous flow rate. Thus parallel pumps are generally used when elevation differences are considerable. Series pumps are used where pipeline elevations are not significantly high.

Example Problem 7.4

One large pump and one small pump are operated in series. The H-Q characteristics of the pumps are defined as follows:

Pump 1

Q, gal/min	0	800	1600	2400	3000
H, ft	2389	2325	2175	1763	1350

Pump 2

Q, gal/min	0	800	1600	2400	3000
H, ft	796	775	725	588	450

(a) Calculate the combined performance of pump 1 and pump 2 in series configuration.
(b) What changes (trimming impellers) must be made to either pump to satisfy the requirement of 2000 ft of head at 2400 gal/min when operated in series?
(c) Can these pumps be configured to operate in parallel?

Solution

(a) Pumps in series have the same flow through each pump and the heads are additive. We can therefore generate the total head produced in series configuration by adding the head of each pump for each flow rate given as follows:

Combined performance of pump 1 and pump 2 in series:

Q, gal/min	0	800	1600	2400	3000
H, ft	3185	3100	2900	2351	1800

(b) It can be seen from the above combined performance that the head generated at 2400 gal/min is 2351 ft. Since the design requirement is 2000 ft at this flow rate, it is clear that the head needs to be reduced by trimming one of the pump impellers to produce the necessary total head. We will proceed as follows.

Let us assume that the smaller pump will not be modified and the impeller of the larger pump (pump 1) will be trimmed to produce the necessary head.

Modified head required of pump 1 = 2000 − 588 = 1412 ft.

Pump 1 produces 1763 ft of head at 2400 gal/min. In order to reduce this head to 1412 ft, we must trim the impeller by approximately

$$(1412/1763)^{1/2} = 0.8949 \text{ or } 89.5\%$$

based on Affinity Laws. This is only approximate and we need to generate a new H-Q curve for the trimmed pump 1 so we can verify that the desired head will be generated at 2000 gal/min. Using the method in Example Problem 7.2 we can generate a new H-Q curve for a trim of 89.5%:

Flow multiplier = 0.8949
Head multiplier = $(0.8949)^2 = 0.8008$

Pump 1 trimmed to 89.5% of present impeller diameter:

Q, gal/min	0	716	1432	2148	2685
H, ft	1913	1862	1742	1412	1081

It can be seen from above trimmed pump performance that the desired head of 1412 ft will be achieved at a flow rate of 2148 gal/min. Therefore, at the lower flow rate of 2000 gal/min, we can estimate by interpolation that the head would be higher than 1412 ft. Hence, slightly more trimming would be required to achieve the design point of 1412 ft at 2000 gal/min. By trial and error we arrive at a pump trim of 87.7% and the resulting pump performance for pump 1 at 87.7% trim is as follows:

Pump 1 trimmed to 87.7% of present impeller diameter:

Q, gal/min	0	702	1403	2105	2631
H, ft	1837	1788	1673	1356	1038

By plotting this curve, we can verify that the required head of 1412 ft will be achieved at 2000 gal/min.

(c) To operate satisfactorily in a parallel configuration, the two pumps must have a common range of heads so that at each common head, the corresponding flow rates can be added to determine the combined performance. Pump 1 and pump 2 are mismatched for parallel operation. Therefore, they cannot be operated in parallel.

7.10 Pump Head Curve Versus System Head Curve

In Chapter 5 we discussed the development of pipeline system head curves. In this section we will see how the system head curve together with the pump head curve can predict the operating point (flow rate Q – head H) on the pump curve.

Since the system head curve for the pipeline is a graphic representation of the pressure required to pump a product through the pipeline at various flow rates (increasing pressure with increasing flow rate) and the pump H-Q curve shows the pump head available at various flow rates, when the head requirements of the pipeline match the available pump head we have a point of intersection of the system head curve with the pump head curve as shown in Figure 7.12. This is the operating point for this pipe-pump combination.

The point of intersection of the pump head curve and the system head curve for diesel (point A) indicates the operating point for this pipeline

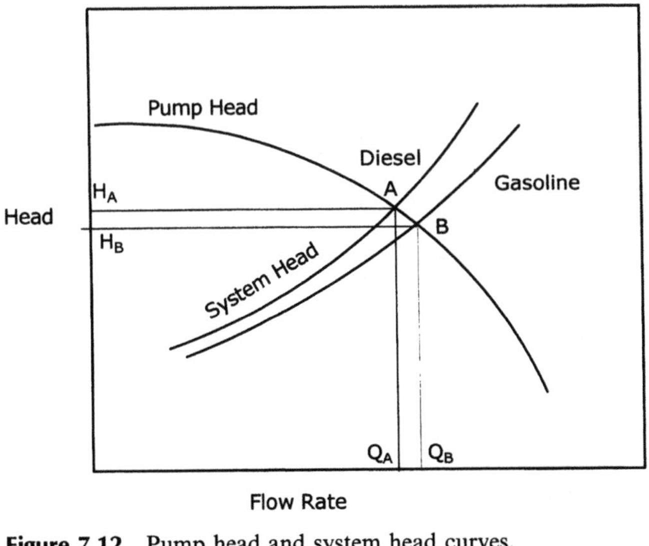

Figure 7.12 Pump head and system head curves.

with diesel. Similarly, if gasoline were pumped through this pipeline, the corresponding operating point (point B) is as shown in Figure 7.12. Therefore, with 100% diesel in the pipeline, the flow rate would be Q_A and the corresponding pump head would be H_A as shown. Similarly, with 100% gasoline in the pipeline, the flow rate would be Q_B and the corresponding pump head would be H_B as shown in Figure 7.12. When batching the two products, a certain proportion of the pipeline will be filled with diesel and the rest will be filled with gasoline. The flow rate will then be at some point between Q_A and Q_B, since a new system head curve located between the diesel curve and the gasoline curve will dictate the operating point.

If we had plotted the system head curve in psi instead of ft of liquid head, the Y-axis will be the pressure required (psi). The pump head curve also needs to be converted to psi versus flow rate by using Equation (3.7):

$$\text{Psi} = \text{Head} \times \text{Sg}/2.31$$

Therefore, using the same pump we will have two separate H-Q curves for diesel and gasoline, due to different specific gravities. Such a situation is shown in Figure 7.13. Even though the pump develops the same head (in ft) with diesel or gasoline (or water), the pressure generated in psi will be different, and hence the two separate H-Q curves, as indicated in Figure 7.13.

In a batched pipeline, the operating point moves from D to A, A to B, B to C, and C to D as shown in Figure 7.13. We start off with 100% diesel in

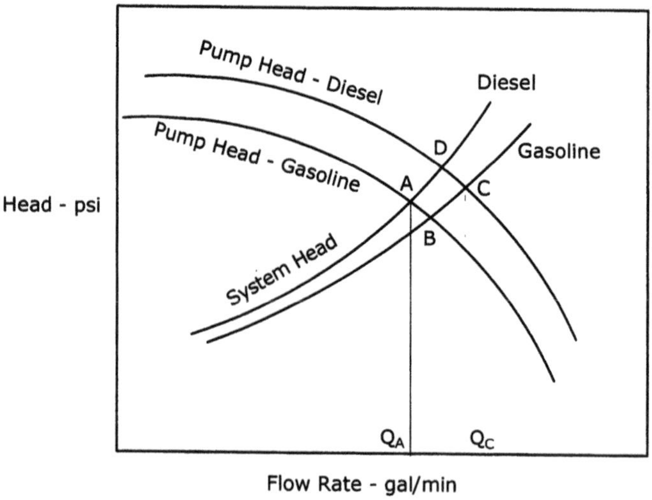

Figure 7.13 Pump and system curve: batching.

the pipeline and the pump. This is point D in the figure. When gasoline enters the pump, with diesel in the pipeline, the operating point moves from D to A. Then, as the gasoline batch enters the pipeline, the system head curve moves to the right until it reaches point B representing the operating point with 100% gasoline in the pipeline and gasoline in the pump. As diesel reaches the pump and the line is still full of gasoline, the operating point moves to C, where we have diesel in the pump and the pipeline filled with gasoline. Finally, as the diesel batch enters the pipeline, the operating point moves towards D. At D we have completed the cycle and both the pump and the pipeline are filled with diesel. Thus it is seen that in a batched operation the flow rates vary between Q_A and Q_C as in Figure 7.13.

7.11 Multiple Pumps Versus System Head Curve

When two or more pumps are operated in parallel on a pipeline system, we saw how the pump head curves added the flow rates at the same head to create the combined pump performance curve. Similarly, with series pumps, the heads are added up for the same flow rate resulting in the combined pump head curve.

Figure 7.14 illustrates the pipeline system head curve superimposed on the pump head curves to show the operating point with one pump, two

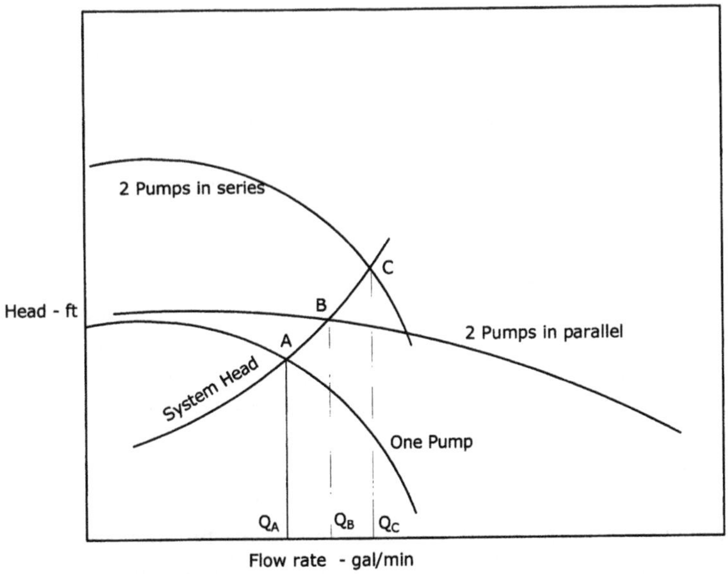

Figure 7.14 Multiple pumps and system head curve.

pumps in series and the same two pumps in parallel configurations. The operating points are shown as A, C, and B with flow rates of Q_A, Q_C, and Q_B, respectively.

In certain pipeline systems, depending upon the flow requirements, we may be able to obtain higher throughput by switching from a series pump configuration to a parallel pump configuration. From Figure 7.14 it can be seen that a steep system head curve would favor pumps in series, while a relatively flat system head curve is associated with the operation of parallel pumps.

7.12 NPSH Required Versus NPSH Available

As the pressure on the suction side of a pump is reduced to a value below the vapor pressure of the liquid being pumped, flashing can occur. The liquid vaporizes and the pump is starved of liquid. At this point the pump is said to cavitate due to insufficient liquid volume and pressure. The vapor can damage the pump impeller, further reducing its ability to pump. To avoid vaporization of liquid, we must provide adequate positive pressure at the pump suction that is greater than the liquid vapor pressure.

NPSH for a centrifugal pump is defined as the net positive suction head required at the pump impeller suction to prevent pump cavitation at any flow rate. Cavitation will damage the pump impeller and render it useless. NPSH represents the resultant positive pressure at the pump suction. In this section, we will analyze a piping configuration from a storage tank to a pump suction, to calculate the available NPSH and compare it with the NPSH required by the pump vendor's performance curve. The NPSH available will be calculated by taking into account any positive tank head, including atmospheric pressure, and subtracting the pressure drop due to friction in the suction piping and the liquid vapor pressure at the pumping temperature. The resulting value of NPSH for this piping configuration will represent the net pressure of the liquid at pump suction, above its vapor pressure. The value calculated must be more than the NPSH specified by the pump vendor at the particular flow rate.

Before we calculate the NPSH available in a typical pump-piping configuration, let us analyze the piping geometry associated with a pump taking suction from a tank and delivering liquid to another tank as shown in Figure 7.15.

The vertical distance from the liquid level on the suction side of the pump center line is defined as the static suction head. More correctly, it is the static suction lift (H_S) when the center line of the pump is above that of the liquid supply level as depicted in Figure 7.15. If the liquid supply level is higher than the pump center line, it is called the static suction head

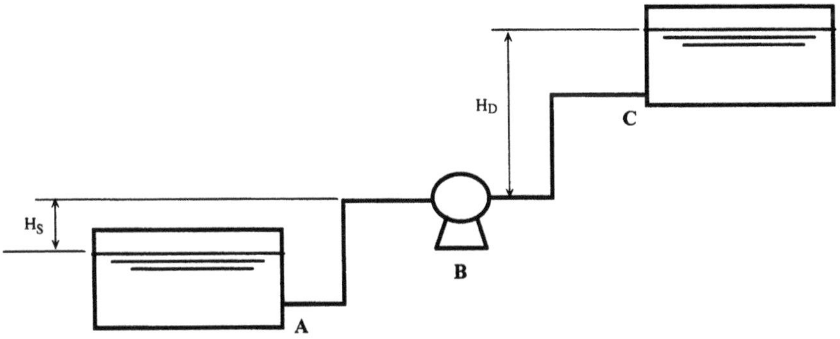

Figure 7.15 Centrifugal pump: suction and discharge heads.

on the pump. Similarly, the vertical distance from the pump center line to the liquid level on the delivery side is called the static discharge head (H_D) as shown in Figure 7.15.

The total static head on a pump is defined as the sum of the static suction head and the static discharge head. It represents the vertical distance between the liquid supply level and the liquid discharge level. The static suction head, static discharge head, and the total static head on a pump are all measured in feet of liquid, or meters of liquid in the SI system.

The friction head, measured in feet of liquid pumped, represents the pressure drop due to friction in both suction and discharge piping. It represents the pressure required to overcome the frictional resistance of all piping, fittings, and valves on the suction side and discharge side of the pump as shown in Figure 7.15. Reviewing the piping system shown in Figure 7.15, it is seen that there are three sections of straight piping and two pipe elbows on the suction side between the liquid supply (A) and the center line of the pump. In addition, there would be at least two valves, one at the tank and the other at the inlet to the pump suction. An entrance loss will be added to account for the pipe entrance at the tank.

Similarly, on the discharge side of the pump between the center line of the pump (B) and the tank delivery point (C) there are three straight sections of pipe and two pipe elbows along with two valves. On the discharge of the pump there would also be a check valve to prevent reverse flow through the pump. An exit loss at the tank entry will also be added to account for the delivery pipe.

On the suction side of the pump, the available suction head H_S will be reduced by the friction loss in the suction piping. This net suction head on the pump will be the available suction head at the pump center line.

On the discharge side the discharge head H_D will be increased by the friction loss in the discharge piping. This is the net discharge head on the pump.

Mathematically,

$$\text{Suction head} = H_S - H_{fs} \tag{7.10}$$

$$\text{Discharge head} = H_D + H_{fd} \tag{7.11}$$

where

H_{fs} = Friction loss in suction piping
H_{fd} = Friction loss in discharge piping

If the suction piping is such that there is a suction lift (instead of suction head) the value of H_S in Equation (7.10) will be negative. Thus a 20 ft static suction lift combined with a suction piping loss of 2 ft will actually result in an overall suction lift of 22 ft. The discharge head, on the other hand, will be the sum of the discharge head and friction loss in the discharge piping. Assuming a 30 ft discharge head and 5 ft friction loss, the total discharge head will be $30+5=35$ ft. In this example, the total head of the pump is $22+35=57$ ft.

Example Problem 7.5

A centrifugal pump is used to pump a liquid from a storage tank through 500 ft of suction piping as shown in Figure 7.16.

(a) Calculate the NPSH available at a flow rate of 3000 gal/min.
(b) The pump vendor's data indicate the NPSH required to be 35 ft at 3000 gal/min and 60 ft at 4000 gal/min. Can this piping system handle the higher flow rate without the pump cavitating?
(c) If cavitation is a problem in (b) above, what changes must be made to the piping system to prevent pump cavitation at 4000 gal/min?

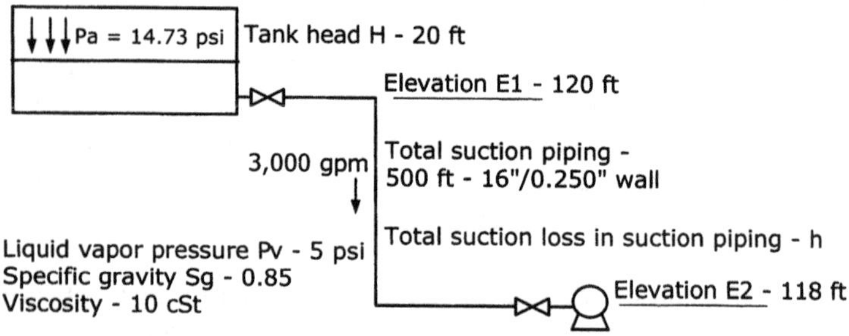

Figure 7.16 NPSH calculation.

Solution

(a) NPSH available in ft of liquid head:

$$(P_a - P_v)(2.31/Sg) + H + E1 - E2 - h \tag{7.12}$$

where

P_a = Atmospheric pressure, psi
P_v = Liquid vapor pressure at the flowing temperature, psi
Sg = Liquid specific gravity
H = Tank head, ft
E1 = Elevation of tank bottom, ft
E2 = Elevation of pump suction, ft
h = Friction loss in suction piping, ft

All terms in Equation (7.12) are known except for the suction piping loss h.

The suction piping loss h is calculated at the given flow rate of 3000 gal/min, considering 500 ft of 16 in. piping, pipe fitting, valves, etc., in the given piping configuration. The total equivalent length of 16 in. pipe, including two gate valves and two elbows, is:

$$\text{Equivalent length of 16 in. pipe} = 500\text{ ft} + 2 \times 8 \times (16/12) + 2 \times 30 \times (16/12)$$

using an L/D ratio of 8 for the gate valves and 30 for each 90° elbow (from Table A.10, Appendix A). Therefore

$$L_e = 500 + 21.33 + 80 = 601.33 \text{ ft}$$

Using the Colebrook-White equation and assuming a pipe roughness of 0.002 in., we calculate the pressure drop at 3000 gal/min as

$$P_m = 12.77 \text{ psi/mile}$$

Therefore

$$h = 12.77 \times 2.31 \times 601.33/(0.85 \times 5280) = 3.95 \text{ ft}$$

Substituting h and other values in Equation (7.12) we get

$$\text{NPSH} = (14.73 - 5) \times 2.31/0.85 + 20 + 120 - 118 - 3.95$$
$$= 44.49 \text{ ft available}$$

(b) At 4000 gal/min

Pressure loss P_m = 21.43 psi/mile

and

$$h = 6.63 \text{ ft}$$

$$\text{Available NPSH at 4000 gal/min} = 44.49 + 3.95 - 6.63 = 41.81 \text{ ft}$$

Since available NPSH is less than the NPSH of 60 ft required by the pump vendor, the pump will cavitate.

(c) The extra head required to prevent cavitation = 60−41.81 = 18 ft.

One solution is to locate the pump suction at an additional 18 ft or more below the tank. Another solution would be to provide a small vertical can pump that can serve as booster for the main pump. This pump will provide the additional head required to prevent cavitation.

7.13 Summary

We have discussed centrifugal pumps and their performance characteristics as they apply to liquid pipeline hydraulics. Other types of pumps such as PD pumps used in injecting liquid into flowing pipelines were briefly covered. Pump performance at different impeller sizes and speeds, based on Affinity Laws, were discussed and illustrated using examples. Trimming impellers or reducing speeds (VSD pumps) to match system pressure requirements was also explored.

The important parameter NPSH was introduced and methods of calculations for typical pump configurations were shown. We also discussed how the water performance curves provided by a pump vendor must be corrected, using the Hydraulic Institute chart, when pumping high-viscosity liquids. The performance of two or more pumps in series or parallel configuration was analyzed using an example. In addition, we demonstrated how the operating point can be determined by the point of intersection of a system head curve and the pump head curve. In batched pipelines the variation of operating point on a pump curve using the system head curves for different products was illustrated.

7.14 Problems

7.14.1 Two pumps are used in series configuration. Pump A develops 1000 ft head at 2200 gal/min and pump B generates 850 ft head at the same flow rate. At the design flow rate of 2200 gal/min, the application requires a head of 1700 ft. The current impeller sizes in both pump are 10 in. diameter.

(a) What needs to be done to prevent pump throttling at the specified flow rate of 2200 gal/min?
(b) Determine the impeller trim size needed for pump A.
(c) If pump A could be driven by a variable-speed drive, at what speed should it be run to match the pipeline system requirement?

7.14.2 A pipeline 40 miles long, 10 in. nominal diameter and 0.250 in. wall thickness is used for gasoline movements. The static head lift is 250 ft from the origin pump station to the delivery point. The delivery pressure is 50 psi and the pump suction pressure is 30 psi. Develop a system head curve for the pipeline for gasoline flow rates up to 2000 gal/min. Use a specific gravity of 0.736 and viscosity of 0.65 cSt at the flowing temperature. Pipe roughness is 0.002 in. Use the Colebrook-White equation.

7.14.3 In Problem 7.14.2 for gasoline movements, the pump used has a head versus capacity and efficiency versus capacity as indicated below:

Q, gal/min	0	400	800	1200	1500
H, ft	3185	3100	2900	2350	1800
E, %	0.0	55.7	78.0	79.3	72.0

(a) What gasoline flow rate will the system operate at with the above pump?
(b) If instead of gasoline, diesel fuel is shipped through the above pipeline on a continuous basis, what throughputs can be expected?
(c) Calculate the motor HP required in cases (a) and (b) above.

7.14.4 A tank farm consists of a supply tank A located at an elevation of 80 ft above mean sea level (MSL) and a delivery tank B at 90 ft above MSL. Two identical pumps are used to transfer liquid from tank A to tank B using interconnect pipe and valves. The pumps are located at an elevation of 50 ft above MSL.

The suction piping from tank A is composed of 120 ft, 14 in. diameter, 0.250 in. wall thickness pipe, six 14 in. 90° elbows, and two 14 in. gate valves. On the discharge side of the pumps, the piping consists of 6000 ft of 12.75 in. diameter,

0.250 in. wall thickness pipe, four 12 in. gate valves, and eight 12 in. 90° elbows. There is a 10 in. check valve on the discharge of each pump. Liquid transferred has a specific gravity of 0.82 and viscosity of 2.5 cSt.

(a) Calculate the total station head for the pumps.
(b) If liquid is transferred at the rate of 2500 gal/min, calculate the friction losses in the suction and discharge piping system.
(c) If one pump is operated while the other is on standby, determine the pump (Q, H) requirement for the above product movement.
(d) Develop a system head curve for the piping system.
(e) What size electric motor (95% efficiency) will be required to pump at the above rate, assuming the pump has an efficiency of 80%?

8

Pump Station Design

In this chapter we will look at some of the significant items in a pump station that pertain to pumps and pipeline hydraulics. We will analyze the various pressures on both the suction and discharge side of the pump and how pressure control is implemented using a control valve downstream of the pumps. The amount of pump throttle pressure due to mismatch between pump head and system curve head will be analyzed and the amount of horsepower wasted will be calculated. The use of variable speed drive (VSD) pumps to eliminate throttle pressure by providing just enough pressure for the specified flow rate is explained. We also perform a simple economic comparison between a VSD pump installation and a constant-speed pump with a control valve.

8.1 Suction Pressure and Discharge Pressure

A typical piping layout within a pump station is as shown in Figure 8.1. The pipeline enters the station boundary at point A, where the station block valve MOV-101 is located. The pipeline leaves the station boundary on the discharge side of the pump station at point B, where the station block valve MOV-102 is located.

Station bypass valves designated as MOV-103 and MOV-104 are used for bypassing the pump station in the event of pump station maintenance

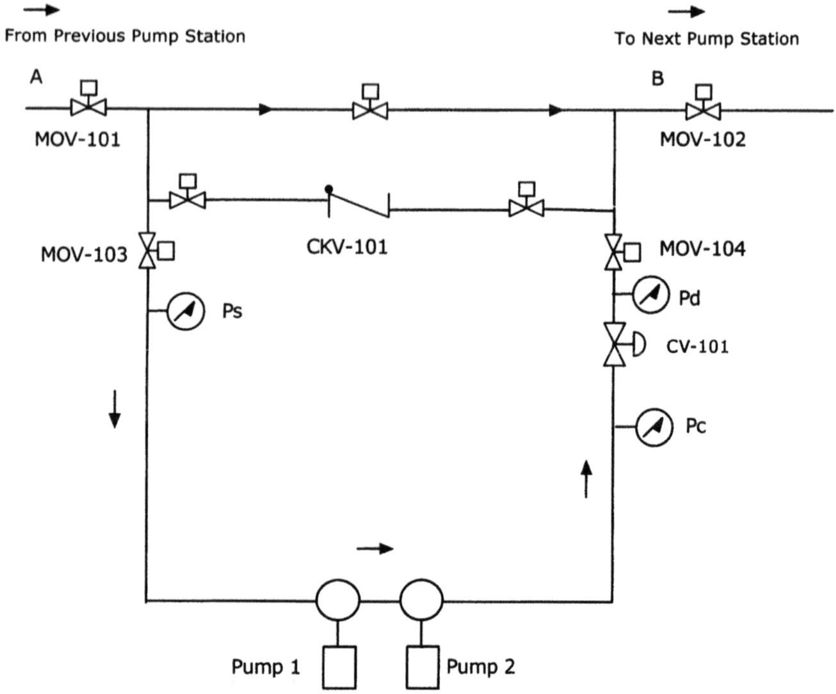

Figure 8.1 Typical pump station layout.

or other reasons when the pump station must be isolated from the pipeline. Along the main pipeline there is located a check valve, CKV-101, that prevents reverse flow through the pipeline. This typical station layout shows two pumps configured in series. Each pump pumps the same flow rate and the total pressure generated is the sum of the pressures developed by each pump. On the suction side of the pump station the pressure is designated as P_s while the discharge pressure on the pipeline side is designated as P_d. With constant-speed motor-driven pumps, there is always a control valve on the discharge side of the pump station, shown as CV-101 in Figure 8.1. This control valve controls the pressure to the required value P_d by creating a pressure drop across it between the pump discharge pressure P_c and the station discharge pressure P_d. Since the pressure within the case of the second pump represents the sum of the suction pressure and the total pressure generated by both pumps, it is referred to as the case pressure P_c.

If the pump is driven by a VSD motor or an engine, the control valve is not needed as the pump may be slowed down or speeded up as required to generate the exact pressure P_d. In such a situation, the case pressure will equal the station discharge pressure P_d.

In addition to the valves shown, there will be additional valves on the suction and discharge of the pumps. Also, not shown in Figure 8.1, is a check valve located immediately after the pump discharge that prevents reverse flow through the pumps.

8.2 Control Pressure and Throttle Pressure

Mathematically, if ΔP_1 and ΔP_2 represent the differential head produced by pump 1 and pump 2 in series, we can write

$$P_c = P_s + \Delta P_1 + \Delta P_2 \tag{8.1}$$

where

P_c = Case pressure in pump 2 or upstream pressure at control valve

The pressure throttled across the control valve is defined as

$$P_{thr} = P_c - P_d \tag{8.2}$$

where

P_{thr} = Control valve throttle pressure

P_d = Pump station discharge pressure

The throttle pressure represents the mismatch that exists between the pump and the system pressure requirements at a particular flow rate. P_d is the pressure at the pump station discharge needed to transport liquid to the next pump station or delivery terminus based on pipe length, diameter, elevation profile, and liquid properties. The case pressure P_c, on the other hand, is the available pressure due to the pumps. If the pumps were driven by variable-speed motors, P_c would exactly match P_d and there would be no throttle pressure. In other words, the control valve would be wide open and obviously would not be needed in operation. The case pressure is also referred to as control pressure since it is the pressure upstream of the control valve. The control valve also functions as a means for protecting the discharge piping. The throttle pressure represents unused pressure developed by the pump and hence results in wasted horsepower (HP) and money. The objective should be to reduce the amount of throttle pressure in any pumping situation.

As mentioned earlier, VSD pumps can speed up or slow down to match pipe pressure requirements. In such situations there is no control valve and therefore the throttle pressure is zero. With VSD pumps there is no HP wasted since the pump case pressure exactly matches the station discharge pressure.

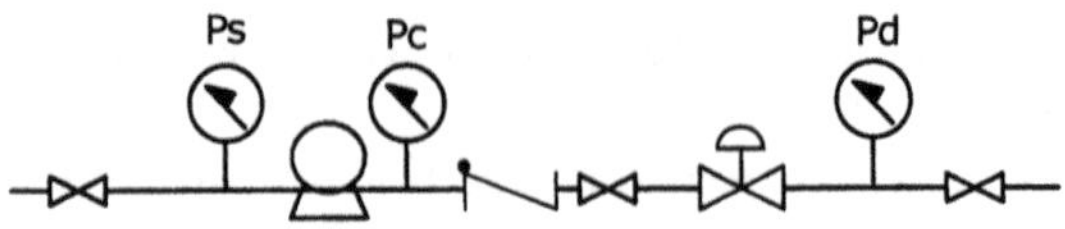

Figure 8.2 Single pump schematic.

A simplified pump station schematic with one pump is shown in Figure 8.2.

Example Problem 8.1

A pump station installation at Corona consists of two pumps configured in series each developing 1500 ft of head at 4000 gal/min. Diesel fuel (Spgr = 0.85, and viscosity = 5.9 cSt) is pumped from Corona to a delivery terminal at Sunnymead, 75 miles away. The required discharge pressure at Corona based on pipe size, pipe length, liquid properties, elevation profile between Corona and Sunnymead, and the required delivery pressure at Sunnymead has been calculated to be 1050 psi. The pump station suction pressure is maintained at 50 psi to prevent pump cavitation. Assume the combined efficiency of both pumps at Corona at the given flow rate to be 82%.

(a) Analyze the pump station pressures and determine the amount of throttle pressure and HP wasted.
(b) If electrical energy costs 8 cents per kWh, estimate the dollars lost in control valve throttling.
(c) What recommendation can you make to improve pipeline operation?

Solution

(a) The total pump pressure developed by two pumps in series is

$$\frac{(1500 + 1500) \times 0.85}{2.31} = 1104 \text{ psi, rounded off}$$

Adding the 50 psi suction pressure gives a control pressure upstream of the control valve of

$$1104 + 50 = 1154 \text{ psi}$$

This is also the case pressure in the second pump. Since the station discharge pressure is 1050 psi, the control valve throttle pressure is

$$\text{Throttle pressure} = 1154 - 1050 = 104 \text{ psi}$$

The HP wasted due to pump throttling can be calculated as

$$\text{HP wasted} = \frac{\text{gal/min} \times \text{psi}}{1714 \times \text{efficiency}} \qquad \text{using Equation (5.18)}$$

or

$$\text{HP wasted} = \frac{4000 \times 104}{1714 \times 0.82} = 296$$

(b) At 8 cents per kWh electric cost, above wasted HP is equivalent to

$$296 \times 0.746 \times 24 \times 350 \times 0.08 = \$148{,}388 \text{ per year}$$

assuming 350 days of continuous operation per year. This is a substantial loss.

(c) If plans are to continue operating this pump station at the current flow rate on a continuous basis, it would be preferable to trim the pump impellers to eliminate the throttle pressure and hence wasted HP. A spare rotating element for the pump can easily be purchased and installed for about $50,000—only about a third of the money wasted annually in throttling. Therefore, the recommendation is to purchase and install a new trimmed rotating element for one of the pumps that will result in no throttling at 4000 gal/min. The existing larger impeller may be stored as a spare for future use when increased flow rates are anticipated.

8.3 Variable-Speed Pumps

It was mentioned in the preceding section that when pumps are driven by variable-speed motors or gas turbine drives, the control valve is not needed and the pump throttle pressures will be zero. The variable-speed pump will be able to speed up or slow down to match the pipeline pressure requirements at any flow rate, thereby removing the need for a control valve. Of course, depending upon the pump, there will be a minimum and maximum permissible pump speeds.

If there are two or more pumps in series configuration, one of the pumps may be driven by a VSD motor or an engine. With parallel pumps, all pumps will have to be VSD pumps, since parallel configuration requires matching heads at the same flow rate.

In the case of two pumps in series, similar to Example Problem 8.1, we could convert one of the two pumps to be driven by a variable-speed motor. This pump can then slow down to the required speed that would develop just the right amount of head (at 4000 gal/min) which, when added to the 1500 ft developed by the constant-speed pump, would provide exactly the total head needed to match the pipeline system requirement of 1050 psi.

Example Problem 8.2

Use the data in Example Problem 8.1 and consider one of the two pumps driven by a variable-speed electric motor. Calculate the speed at which the VSD pump should be operated to match the pipeline requirements at 4000 gal/min. Assume that the rated speed for the pump at 3560 RPM produces a head of 1500 ft at 4000 gal/min.

Solution

The pipeline pressure requirement is given as 1050 psi. Assuming the VSD pump develops a head of H ft, the total head produced by the two pumps in series is (H + 1500) ft. Since the pump suction pressure is 50 psi, the total pressure on the discharge side of the pumps is

$$(H + 1500) \times 0.85/2.31 + 50 \text{ psi}$$

This must exactly equal the discharge pressure requirement of 1050 psi at the 4000 gal/min flow rate. Therefore, we can state that

$$1050 = 50 + (H + 1500) \times 0.85/2.31$$

From which, we get

$$H = 1218 \text{ ft}$$

This is the head that the VSD pump must develop, compared with 1500 ft for the constant-speed pump.

At the rated speed of 3560 RPM, this pump develops 1500 ft at 4000 gal/min. In order to reduce the head to 1218 ft, the VSD pump has to be slowed down to some speed N RPM. We will calculate this speed using Affinity Laws.

Since head is proportional to the square of the speed,

$$\text{Speed ratio} = N/3560 = (1218/1500)^{1/2} \qquad \text{using Equation (7.8)}$$

Solving for N we get

$$N = 3208 \text{ RPM approximately}$$

By operating the second pump (as a VSD pump) at 3208 RPM in series with the first pump (constant speed) the pump station will produce exactly 1050 psi required to pump the diesel fuel at 4000 gal/min. Therefore, no pump throttling will be required and no pump BHP will be wasted.

8.3.1 VSD Pump Versus Control Valve

In a single pump station pipeline with one pump used to provide the pressure required to pump the liquid, a control valve is used to regulate the pressure for a given flow rate.

Suppose the pipeline from the Essex pump station to the Kent delivery terminal is 120 miles long and is constructed of a 16 in. diameter and 0.250 in. wall thickness pipe with a maximum allowable operating pressure (MAOP) of 1440 psi (Figure 8.3). The pipeline is designed to operate at 1400 psi, pumping 4000 bbl/hr of liquid (specific gravity 0.89 and viscosity 30 cSt at 60°F) on a continuous basis. The delivery pressure required at Kent is 50 psi. If the pump suction pressure at Essex is 50 psi, the required pump differential pressure is 1350 psi. Unless the single pump unit at Essex was specifically selected for this application, the chances are that the Essex pump H-Q curve may indicate the following:

$$Q = 2800 \text{ gal/min} \qquad H = 3800 \text{ ft}$$

The flow rate of 2800 gal/min is the equivalent of 4000 bbl/hr. Converting the head available (3800 ft) into psi, the pump pressure developed is

$$3800 \times 0.89/2.31 = 1464 \text{ psi}$$

On the other hand, if the pump head at Q = 2800 gal/min were only 3200 ft, the pressure developed would only be

$$3200 \times 0.89/2.31 = 1233 \text{ psi}$$

which is inadequate for our application that requires a pressure of 1350 psi to pump the liquid at 4000 bbl/hr.

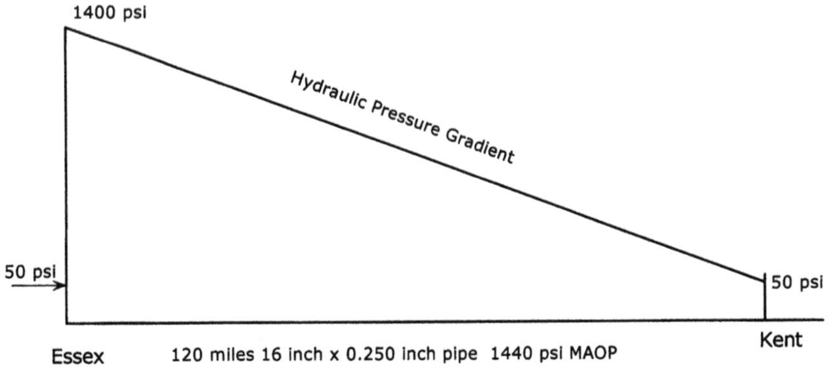

Figure 8.3 Essex to Kent pipeline.

Assuming the first case where the pump is able to develop 3800 ft that corresponds to 1464 psi, we can see that this pressure combined with the 50 psi pump suction pressure would produce a pump discharge pressure (actually the pump case pressure) of

$$1464 + 50 = 1514 \text{ psi}$$

As mentioned earlier, the pipeline pressure limit (MAOP) is 1440 psi; as a result we will over-pressure the pipeline by 74 psi. Obviously, this cannot be tolerated and we must have some means to control the pump pressure to the required 1400 psi pump station discharge. A control valve located downstream of the pump discharge will be used to reduce the discharge pressure to the required pressure of 1400 psi. This is depicted by the modified system curve (2) in Figure 8.4.

The system head curve (1) in Figure 8.4 represents the pressure versus flow rate variation for our pipeline from Essex to Kent. At Q = 2800 gal/min (4000 bbl/hr) flow rate, point C on the pipeline system head curve shows the desired operating point that requires a pipeline discharge pressure of 1400 psi at Essex. Since we have superimposed the pump H-Q curve and pressures are in ft of liquid head, the pressure at C is

$$1400 \times 2.31/0.89 = 3634 \text{ ft}$$

Since the pump suction pressure is 50 psi, we have plotted the pump curve in Figure 8.4 to include this suction head as well. The pump H-Q curve

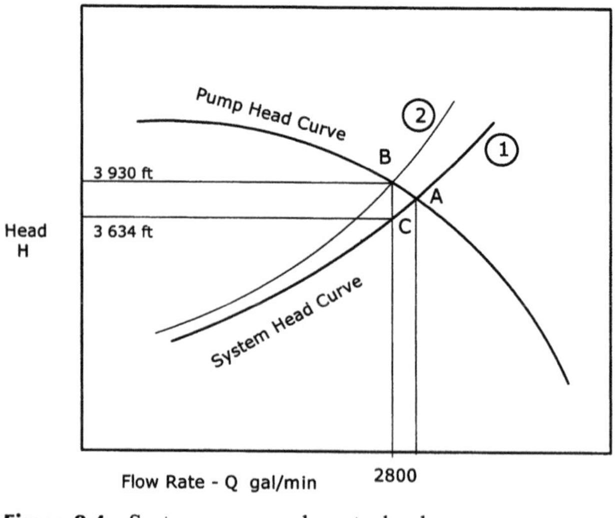

Figure 8.4 System curve and control valve.

shows that at a flow rate of 2800 gal/min, the differential head generated is 3800 ft. Adding the 50 psi suction pressure, point B on the H-Q curve corresponds to

$$3800 + 50 \times 2.31/0.89 = 3930 \text{ ft}$$

We can see from the point of intersection of the pump H-Q curve and the pipeline system curve that the operating point will be at point A, which is a higher flow rate than 2800 gal/min. Also, the pump discharge pressure at A will be at a higher flow rate than that at C. Unless the pipe MAOP is greater than the pressure at A, we cannot allow this pump to operate without pressure control. Thus utilizing the control valve on the pump discharge we will move the operating point from A to B on the pump curve corresponding to 2800 gal/min and a higher head of 3930 ft as calculated above. The control valve will then cause the pressure drop equivalent to the head difference BC calculated as follows:

$$\text{Control valve pressure drop, BC} = 3930 - 3634 = 296 \text{ ft}$$

This pressure drop across the control valve is also called the pump throttle pressure:

$$\text{Throttle pressure} = 296 \times 0.89/2.31 = 114 \text{ psi}$$

We thus conclude that, due to the slightly oversized pump and the pipe MAOP limit, we must utilize a control valve on the pump discharge to limit the pipeline discharge pressure to 1400 psi at the required flow rate of 4000 bbl/hr (2800 gal/min). A hypothetical system head curve, designated as (2), passing through point B on the pump curve, is the artificial system head curve due to the restriction imposed by the control valve.

It must be pointed out that as far as the pump is concerned, the suction pressure, case pressure, discharge pressure, and throttle pressure are as follows

Pump suction pressure = 50 psi
Pump case pressure = 1514 psi
Pump station discharge pressure = 1400 psi
Pump throttle pressure = 114 psi

The above analysis applies to a pump driven by a constant-speed electric motor. If we had a VSD pump that can vary the pump speed from 60% to 100% rated speed and if the rated speed were 3560 RPM, the pump speed range would be

$$3560 \times 0.60 = 2136 \text{ RPM to } 3560 \text{ RPM}$$

Since the pump speed can be varied from 2136 RPM to 3560 RPM, the pump head curve will correspondingly vary according to the centrifugal pump Affinity Laws. Therefore we can find some speed (less than 3560 RPM) at which the pump will generate the required head corresponding to point C in Figure 8.4.

Point C represents a pump differential head of

$$(1400 - 50) \times 2.31/0.89 = 3504 \text{ ft}$$

where 50 psi represents the pump suction pressure. If the given H-Q curve is based on the rated speed of 3560 RPM, using Affinity Laws we calculate the approximate speed required for point C as

$$3560(3504/3800)^{1/2} = 3419 \text{ RPM}$$

If we had the full H-Q curve data at 3560 RPM we could generate the revised H-Q curve at the desired speed of 3419 RPM. Actually, we may have to adjust the speed a little to get the required head corresponding to point C.

From the foregoing analysis we can conclude that use of VSD pumps can provide the right amount of pressure required for a given pump flow rate, thus avoiding the pump throttle pressures (and hence wasted HP) that are common with constant-speed motor-driven pumps with control valves. However, VSD pumps are expensive to install and operate compared with the use of a control valve. A typical control valve installation may cost \$100,000, whereas a VSD may require \$300,000 to \$500,000 incremental cost compared with a constant-speed motor-driven pump. We will have to factor in the increased operating cost of the VSD pump compared to the money lost in wasted HP from control valve throttling.

Let us illustrate this function by considering the Essex to Kent pipeline again. In the first case, we will assume a constant-speed pump at Essex with a control valve. The pump and control valve installation will cost approximately

$$\$(X + 100{,}000)$$

where X represents the cost of the constant-speed motor-driven pump. The annual maintenance cost for the control valve is estimated at \$10,000. Due to pump throttling, let us assume an average pressure drop in the control valve to be 100 psi over a 12 month period at an average pipeline flow rate of 4000 bbl/hr. This is equivalent to a wasted HP of

$$\frac{2800 \times 100}{1714 \times 0.8} = 204.2 \text{ HP}$$

Assuming 80% pump efficiency, at an average electrical cost of 10 cents/kWh and 350 days operation per year, the throttled pressure represents a loss of

$$\$(204.2 \times 0.746 \times 24 \times 350 \times 0.10) = \$127{,}960 \text{ per year}$$

If we now replaced the pump at Essex with a VSD pump consisting of a variable frequency drive (VFD) electric motor with the same pump without the control valve, the capital cost will be approximately

$$\$(X + 500{,}000)$$

where $500,000 has been added to account for the additional switchgear and the VFD motor. The annual maintenance cost for this installation may be as high as $50,000.

Comparing the two alternatives, we can summarize as follows, assuming $X = 250{,}000$ for the pump motor.

Case A: Constant-speed motor-driven pump with control valve

Capital cost $350,000
Annual cost $137,960
(includes HP lost due to control valve)

Case B: VFD motor-driven pump, no control valve

Capital cost $750,000
Annual cost $50,000

It is clear that the VSD pump has higher initial cost but lower annual cost compared with the constant-speed motor/control valve combination.

In this example, we can perform an economic analysis and compare case A with case B considering discounted cash flow. If the interest rate is 8% per year the present value of case A for a 15 year project life is

$$PV_A = 350{,}000 + PV(137{,}960,\ 15,\ 8)$$

where PV(A, N, i) represents the present value of a series of cash flows of $A, each for N years at an annual interest rate of i percent. Therefore

$$PV_A = 350{,}000 + 1{,}180{,}866$$
$$= \$1{,}530{,}866$$

Similarly,

$$PV_B = 750{,}000 + PV(50{,}000,\ 15,\ 8)$$

$$= 750{,}000 + 427{,}974$$

$$= \$1{,}177{,}974$$

It can be seen from the above that the VSD case will be the more economical alternative, since it has a lower present value of investment.

The advantages of VSD pumps include operational flexibility and lower power requirements since no HP is wasted as with a control valve. The disadvantage of VSD pumps include higher capital and operating cost compared with a constant-speed pump with a control valve.

8.4 Summary

This chapter covered the main components found in a typical pump station as they relate to pumps and pipeline hydraulics. The concepts of pump suction pressure, case pressure, pump station discharge pressure, and throttle pressure were introduced and calculations illustrated using examples. The impact of pump throttle pressure and how the control valve is used to protect the discharge piping as well as to provide the necessary pressure for a particular flow rate were explained. We found that the throttle pressure contributes to energy loss and therefore wasted money. The advantages of using VSD pumps to provide just enough pressure for the specified flow, thereby eliminating throttle pressures, were analyzed. Also illustrated was a comparison of using a VSD pump versus a control valve.

8.5 Problems

8.5.1 The origin pump station at Harvard on a 50 mile pipeline consists of two pumps in series, each developing 1450 ft of head at 3000 bbl/hr. Diesel fuel (specific gravity 0.85 and viscosity 5.9 cSt) is pumped from Harvard to a delivery terminal at Banning. The required discharge pressure at Harvard has been calculated to be 995 psi. The pump station suction pressure is maintained at 50 psi. Use a combined pump efficiency of 84% at the given flow rate.

(a) Analyze the pump station pressures and determine the amount of throttle pressure and HP wasted.

(b) If electrical energy costs 10 cents/kWh, estimate the dollars lost in control valve throttling.

8.5.2 If a VSD pump is used in Problem 8.5.1 above, determine the speed at which the pump should be run to maintain the flow rate of 3000 bbl/hr. The rated pump speed is 3560 RPM.

8.5.3 A water pipeline consists of three pump stations, each having one constant-speed motor-driven pump of 2000 HP. The analysis of current operations shows that the pipeline has been operating at a fairly constant flow rate of 3000 gal/min, and the pump throttle pressures are as follows:

Pump station 2: 75 psi
Pump station 3: 85 psi

Pump station 1 has zero throttle pressure. Trimming the pump impellers to the correct size so as to eliminate wasted HP from pump throttle is estimated to cost \$50,000 per pump station. Retrofitting each pump station with VFD motor-driven pumps is estimated to cost \$500,000 per pump station. The operating cost for the VFD units is \$50,000 per site.

(a) Determine the economics of installing trimmed impellers and the payment period based on 8% interest rate. Assume 80% pump efficiency for HP calculations and an electricity cost of 10 cents/kWh.
(b) If it was decided not to trim the pump impellers, compare the VFD motor option with the constant-speed pump option. Consider the control valve annual operating cost to be \$5000 per site.

8.5.4 A refined products pipeline is used for batched operations with gasoline, kerosene and diesel fuel. The pipeline is 80 miles long and is 14 in. in diameter, 0.250 in. wall thickness. The MAOP of the pipeline is 1200 psi.

(a) Determine the maximum throughput possible with one pump station for each product considering each product pumped alone with no batching. Assume a flat elevation profile with 50 psi pipe delivery pressure. For liquid properties use the following:

Product	Specific gravity	Viscosity (cSt)
Gasoline	0.74	0.6
Kerosene	0.82	2.0
Diesel	0.85	5.5

(b) Investigate the use of VSD pumps to provide the required operational control and flexibility when this pipeline is operated in a batched mode. Pick a typical pump curve for the origin pump station to first satisfy the need to maximize gasoline throughput. Having based pump requirements on gasoline, determine the speed variants required for the other two products without exceeding the MAOP.

(c) Estimate the annual power cost for each product based on an electricity cost of 10 cents/kWh.

(d) How would the operating scenario change if VSD pumps were not used? Estimate the cost of HP wasted in pump throttling with the control valve.

9

Thermal Hydraulics

Thermal hydraulics takes into account the temperature variation of a liquid as it flows through the pipeline. This is in contrast to isothermal hydraulics, where there is no significant temperature variation in the liquid. In previous chapters we have concentrated on water pipelines, refined petroleum products (gasoline, diesel, etc.) and other light crude oil pipelines where the liquid temperature was close to ambient temperature. In many cases where heavy crude oil and other liquids of high viscosity have to be pumped, the liquid is heated to some temperature (such as 150°F to 180°F) prior to being pumped through the pipeline. In this chapter we explore how calculations are performed in thermal hydraulics.

9.1 Temperature-Dependent Flow

In the preceding chapters we concentrated on steady-state liquid flow in pipelines without paying much attention to temperature variations along the pipeline. We assumed the liquid entered the pipeline inlet at some temperature such as 70°F. The liquid properties such as specific gravity and viscosity at the inlet temperature were used to calculate the Reynolds number and friction factor and finally the pressure drop due to friction. Similarly, we also used the specific gravity at inlet temperature to calculate the elevation head based on the pipeline topography. In all cases the liquid properties were considered at some constant flowing temperature. These calculations are therefore based on isothermal (constant-temperature) flow.

The above may be valid in most cases in which the liquid transported, such as water, gasoline, diesel, or light crude oil, is at ambient temperature. As the liquid flows through the pipeline, heat may be transferred to or from the liquid from the surrounding soil (buried pipeline) or the ambient air (above-ground pipeline). Significant changes in liquid temperatures due to heat transfer with the surroundings will affect liquid properties such as specific gravity and viscosity. This in turn will affect pressure drop calculations. So far we have ignored this heat transfer effect, assuming minimal temperature variations along the pipeline. However, there are instances when the liquid has to be heated to a much higher temperature than ambient conditions to reduce the viscosity and make it flow easily. Pumping a higher-viscosity liquid that is heated will also require less pump horsepower.

For example, a high-viscosity crude oil (200 cSt to 800 cSt or more at 60°F) may be heated to 160°F before it is pumped into the pipeline. This high-temperature liquid loses heat to the surrounding soil as it flows through the pipeline by conduction of heat from the interior of the pipe to the soil through the pipe wall. The ambient soil temperature may be 40°F to 50°F during the winter and 60°F to 80°F during the summer. Therefore, a considerable temperature difference exists between the hot liquid in the pipe and the surrounding soil.

The temperature difference of about 120°F in winter and 100°F during summer, will cause significant heat transfer between the crude oil and surrounding soil. This will result in a temperature drop of the liquid and variation in liquid specific gravity and viscosity as it flows through the pipeline. Therefore, in such instances we will be wrong in assuming a constant flowing temperature to calculate pressure drop as we do in isothermal flow. Such a heated liquid pipeline may be bare or insulated. In this chapter we will study the effect of temperature variation and friction loss along the pipeline, known as thermal hydraulics.

Consider a 20 in. buried pipeline transporting 8000 bbl/hr of a heavy crude oil that enters the pipeline at an inlet temperature of 160°F. Assume that the liquid temperature has dropped to 124°F at a location 50 miles from the pipeline inlet. Suppose the crude oil properties at 160°F inlet conditions and at 124°F at milepost 50 are as follows:

Temperature (°F)	Specific gravity	Viscosity (cSt)
160	0.9179	40.55
124	0.9306	103.69

Using the 160°F inlet temperature we calculate the frictional pressure drop at inlet conditions to be 7.6 psi/mile. At the temperature conditions at milepost 50, using the given liquid properties we find that the frictional pressure drop has increased to 23.97 psi/mile. Thus, in a 50 mile section of pipe the pressure drop due to friction varies from 7.6 to 23.97 psi/mile as shown in Figure 9.1.

We could use an average value of the pressure drop per mile to calculate the total frictional pressure drop in the first 50 miles of the pipe. However, this will be a very rough estimate. A better approach would be to subdivide the 50 mile section of the pipeline into smaller segments 5 or 10 miles long and to compute the pressure drop per mile for each segment. We will then add the individual pressure drops for each 5 or 10 mile segment to get the total frictional pressure drop in the 50 mile length of pipeline. Of course, this assumes that we have available to us the temperature of the liquid at 5 or 10 mile increments up to milepost 50. The liquid properties and pressure drop due to friction can then be calculated at the boundaries of each 5 or 10 mile segment and the temperature gradient plotted as illustrated in Figure 9.2.

How do we obtain the temperature variation along the pipeline so we can calculate the liquid properties at each temperature and then calculate the pressure drop due to friction? This represents the most complicated aspect of thermal hydraulic analysis. Several approaches have been put forth for calculating the temperature variation in a pipeline transporting heated liquid. We must consider soil temperatures along the pipeline, thermal conductivity of pipe material, pipe insulation (if any), thermal conductivity

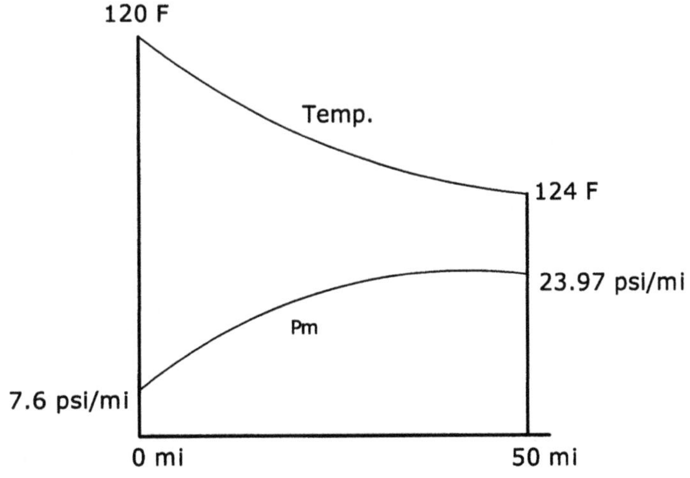

Figure 9.1 Temperature and pressure drop.

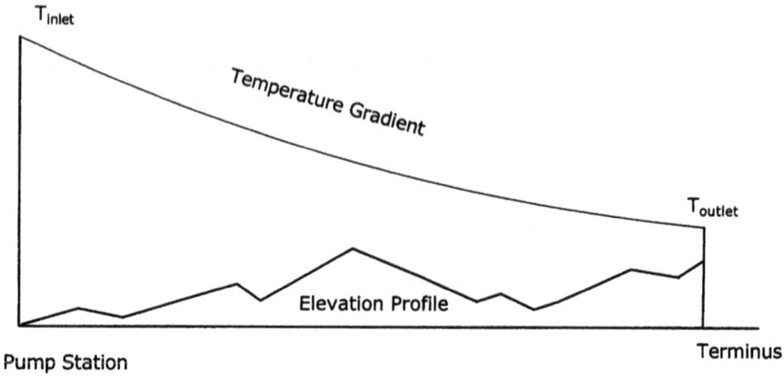

Figure 9.2 Thermal temperature gradient.

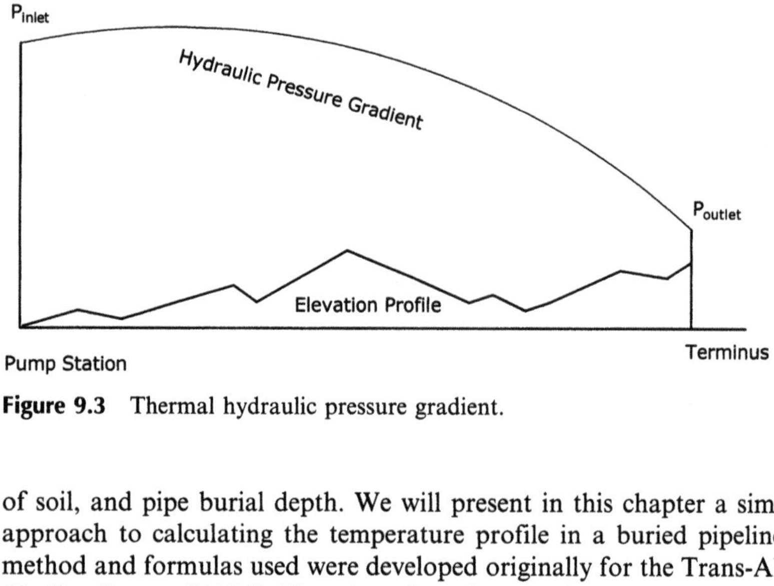

Figure 9.3 Thermal hydraulic pressure gradient.

of soil, and pipe burial depth. We will present in this chapter a simplified approach to calculating the temperature profile in a buried pipeline. The method and formulas used were developed originally for the Trans-Alaskan Pipeline System (TAPS). These have been found to be quite accurate over the range of temperatures and pressures encountered in heated liquid pipelines today.

The hydraulic gradient showing the pressure profile in a heated liquid pipeline is illustrated in Figure 9.3. Note the curved shape of the gradient, compared with the straight-line gradient in isothermal flow (such as Figures 6.2 or 8.3).

An example will illustrate the use of these formulas. More accurate methods include using a computer software program that will subdivide the pipeline into small segments and compute the heat balance and pressure drop calculations, to develop the pressure and temperature profile for the entire pipeline. One such commercially available program is LIQTHERM developed by SYSTEK (www.systek.us).

9.2 Formulas for Thermal Hydraulics

9.2.1 Thermal Conductivity

Thermal conductivity is the property used in heat conduction through a solid. In English units it is measured in Btu/hr/ft/°F. In SI units thermal conductivity is expressed in W/m/°C.

For heat transfer through a solid of area A and thickness dx, with a temperature difference dT, the formula in English units is:

$$H = K(A)(dT/dx) \tag{9.1}$$

where

H = Heat flux perpendicular to the surface area, Btu/hr
K = Thermal conductivity of solid, Btu/hr/ft/°F
A = Area of heat flux, ft^2
dx = Thickness of solid, ft
dT = Temperature difference across the solid, °F

The term dT/dx represents the temperature gradient in °F/ft. Equation (9.1) is also known as the Fourier heat conduction formula.

It can be seen from Equation (9.1) that the thermal conductivity of a material is numerically equal to the amount of heat transferred across a unit area of the solid material with unit thickness, when the temperature difference between the two faces of the solid is maintained at 1 degree. The thermal conductivity for steel pipe and soil are as follows:

K for steel pipe = 29 Btu/hr/ft/°F

K for soil = 0.2 to 0.8 Btu/hr/ft/°F

Sometimes heated liquid pipelines are insulated on the outside with an insulating material. The K value for insulation may range from 0.01 to 0.05 Btu/hr/ft/°F.

In SI units, Equation (9.1) becomes

$$H = K(A)(dT/dx) \tag{9.2}$$

where

H = Heat flux, W
K = Thermal conductivity of solid, W/m/°C
A = Area of heat flux, m^2
dx = Thickness of solid, m

dT = Temperature difference across the solid, °C

In SI units, the thermal conductivity for steel pipe, soil and insulation are as follows:

K for steel pipe = 50.19 W/m/°C

K for soil = 0.35 to1.4 W/m/°C

K for insulation = 0.02 to 0.09 W/m/°C

As an example, consider heat transfer across a flat steel plate of thickness 8 in. and area 100 ft^2 where the temperature difference across the plate thickness is 20°F. From Equation (9.1)

$$\text{Heat transfer} = 29 \times (100) \times 20 \times 12/8 = 87{,}000 \text{ Btu/hr}$$

9.2.2 Overall Heat Transfer Coefficient

The overall heat transfer coefficient is also used in heat flux calculations. Equation (9.1) for heat flux can be written in terms of overall heat transfer coefficient as follows:

$$H = U(A)(dT) \tag{9.3}$$

where

U = Overall heat transfer coefficient, Btu/hr/ft^2/°F

Other symbols in Equation (9.3) are the same as in Equation (9.1).

In SI units, Equation (9.3) becomes

$$H = U(A)(dT) \tag{9.4}$$

where

U = Overall heat transfer coefficient, W/m^2/°C

Other symbols in Equation (9.4) are the same as in Equation (9.2). The value of U may range from 0.3 to 0.6 Btu/hr/ft^2/°F in English units and 1.7 to 3.4 W/m^2/°C in SI units.

When analyzing heat transfer between the liquid in a buried pipeline and the outside soil, we consider flow of heat through the pipe wall and pipe insulation (if any) to the soil. If U represents the overall heat transfer coefficient, we can write from Equation (9.3)

$$H = U(A)(T_L - T_S) \tag{9.5}$$

where

A = Area of pipe under consideration, ft^2

T_L = Liquid temperature, °F

T_S = Soil temperature, °F
U = Overall heat transfer coefficient, Btu/hr/ft^2/°F

Since we are dealing with temperature variation along the pipeline length, we must consider a small section of pipeline at a time when applying above Equation (9.5) for heat transfer.

For example, consider a 100 ft length of 16 in. pipe carrying a heated liquid at 150°F. If the outside soil temperature is 70°F and the overall heat transfer coefficient

$$U = 0.5 \text{ Btu/hr/ft}^2/°\text{F}$$

then we can calculate the heat transfer using Equation (9.5) as follows:

$$H = 0.5\,A\,(150 - 70)$$

where A is the area through which heat flux occurs:

$$A = \pi \times (16/12) \times 100 = 419 \text{ ft}^2$$

Therefore

$$H = 0.5 \times 419 \times 80 = 16{,}760 \text{ Btu/hr}$$

9.2.3 Heat Balance

The pipeline is subdivided and for each segment the heat content balance is computed as follows:

$$H_{in} - \Delta H + H_w = H_{out} \tag{9.6}$$

where

H_{in} = Heat content entering line segment, Btu/hr
ΔH = Heat transferred from line segment to surrounding medium (soil or air), Btu/hr
H_w = Heat content from frictional work, Btu/hr
H_{out} = Heat content leaving line segment, Btu/hr

In the above we have included the effect of frictional heating in the term H_w. With viscous liquids the effect of friction is to create additional heat which would raise the liquid temperature. Therefore in thermal hydraulic analysis frictional heating is included to improve the calculation accuracy.

In SI units, Equation (9.6) will be the same, with each term expressed in watts instead of Btu/hr.

The heat balance Equation (9.6) forms the basis for computing the outlet temperature of the liquid in a segment, starting with its inlet temperature and taking into account the heat loss (or gain) with the surroundings and frictional heating. In the following sections we will formulate the method of calculating each term in Equation (9.6).

9.2.4 Logarithmic Mean Temperature Difference (LMTD)

In heat transfer calculations, due to varying temperatures it is customary to use a slightly different concept of temperature difference called the logarithmic mean temperature difference (LMTD). The LMTD between the liquid in the pipeline and the surrounding medium is calculated as follows:

Consider a pipeline segment of length Δx with liquid temperatures T_1 at the upstream end and T_2 at the downstream end of the segment. If T_s represents the average soil temperature (or ambient air temperature for an above-ground pipeline) surrounding this pipe segment, the logarithmic mean temperature of the pipe (T_m) segment is calculated as follows:

$$T_m - T_S = \frac{(T_1 - T_S) - (T_2 - T_S)}{\text{Log}_e[(T_1 - T_S)/(T_2 - T_S)]} \tag{9.7}$$

where

T_m = Logarithmic mean temperature of pipe segment, °F
T_1 = Temperature of liquid entering pipe segment, °F
T_2 = Temperature of liquid leaving pipe segment, °F
T_S = Sink temperature (soil or surrounding medium), °F

In SI units, Equation (9.7) will be the same, with all temperatures expressed in °C instead of °F. For example, if the average soil temperature is 60°F and the temperatures of the pipe segments upstream and downstream are 160°F and 150°F, respectively, the logarithmic mean temperature of the pipe segment is:

$$T_m = 60 + \frac{(160 - 60) - (150 - 60)}{\text{Log}_e[(160 - 60)/(150 - 60)]} = 60 + 94.88 = 154.88°\text{F}$$

We have thus calculated the logarithmic mean temperature of the pipe segment to be 154.88°F. If we had used a simple arithmetic average we would get the following for the mean temperature of the pipe segment:

Arithmetic mean temperature = (160 + 150)/2 = 155°F

This is not too far off the logarithmic mean temperature T_m calculated above. It can be seen that the logarithmic mean temperature approach gives a slightly more accurate representation of the average liquid temperature in the pipe segment. Note that the use of natural logarithm in Equation (9.7) signifies an exponential decay of the liquid temperature in the pipeline segment. In this example, the LMTD for the pipe segment is

$$\text{LMTD} = 154.88 - 60 = 94.88°\text{F}$$

If we assume an overall heat transfer coefficient $U = 0.5$ Btu/hr/ft^2/°F, we can estimate the heat flux from this pipe segment to the surrounding soil using Equation (9.4) as follows:

$$\text{Heat flux} = 0.5 \times 1 \times 94.88 = 47.44 \text{ Btu/hr per ft}^2 \text{ of pipe area}$$

9.2.5 Heat Entering and Leaving Pipe Segment

The heat content of the liquid entering and leaving a pipe segment is calculated using the mass flow rate of the liquid, its specific heat and the temperatures at the inlet and outlet of the segment. The heat content of the liquid entering the pipe segment is calculated from

$$H_{in} = w(C_{pi})(T_1) \tag{9.8}$$

The heat content of the liquid leaving the pipe segment is calculated from

$$H_{out} = w(C_{po})(T_2) \tag{9.9}$$

where

H_{in} = Heat content of liquid entering pipe segment, Btu/hr
H_{out} = Heat content of liquid leaving pipe segment, Btu/hr
C_{pi} = Specific heat of liquid at inlet, Btu/lb/°F
C_{po} = Specific heat of liquid at outlet, Btu/lb/°F
w = Liquid flow rate, lb/hr
T_1 = Temperature of liquid entering pipe segment, °F
T_2 = Temperature of liquid leaving pipe segment, °F

The specific heat C_p of most liquids ranges between 0.4 and 0.5 Btu/lb/°F (0.84 and 2.09 kJ/kg/°C) and increases with liquid temperature. For petroleum fluids C_p can be calculated if the specific gravity or API gravity and temperatures are known.

In SI units Equations (9.8) and (9.9) become

$$H_{in} = w(C_{pi})(T_1) \tag{9.10}$$

$$H_{out} = w(C_{po})(T_2) \tag{9.11}$$

where

H_{in} = Heat content of liquid entering pipe segment, J/s (W)
H_{out} = Heat content of liquid leaving pipe segment, J/s (W)

C_{pi} = Specific heat of liquid at inlet, kJ/kg/°C
C_{po} = Specific heat of liquid at outlet, kJ/kg/°C
w = Liquid flow rate, kg/s
T_1 = Temperature of liquid entering pipe segment, °C
T_2 = Temperature of liquid leaving pipe segment, °C

9.2.6 Heat Transfer: Buried Pipeline

Consider a buried pipeline, with insulation, that transports a heated liquid. If the pipeline is divided into segments of length L we can calculate the heat transfer between the liquid and the surrounding medium using the following equations:

In English units

$$H_b = 6.28(L)(T_m - T_S)/(\text{Parm1} + \text{Parm2}) \quad (9.12)$$

$$\text{Parm1} = (1/K_{ins})\text{Log}_e(R_i/R_p) \quad (9.13)$$

$$\text{Parm 2} = (1/K_s)\text{Log}_e[2S/D + ((2S/D)^2 - 1)^{1/2}] \quad (9.14)$$

where

H_b = Heat transfer, Btu/hr
T_m = Log mean temperature of pipe segment, °F
T_S = Ambient soil temperature, °F
L = Pipe segment length, ft
R_i = Pipe insulation outer radius, ft
R_p = Pipe wall outer radius, ft
K_{ins} = Thermal conductivity of insulation, Btu/hr/ft/°F
K_s = Thermal conductivity of soil, Btu/hr/ft/°F
S = Depth of cover (pipe burial depth) to pipe centerline, ft
D = Pipe outside diameter, ft

Parm1 and Parm 2 are intermediate values that depend on parameters indicated.

In SI units, Equations (9.12), (9.13), and (9.14) become

$$H_b = 6.28(L)(T_m - T_s)/(\text{Parm1} + \text{Parm 2}) \quad (9.15)$$

$$\text{Parm1} = (1/K_{ins})\text{Log}_e(R_i/R_p) \quad (9.16)$$

$$\text{Parm2} = (1/K_s)\text{Log}_e[2S/D + ((2S/D)^2 - 1)^{1/2}] \quad (9.17)$$

where

H_b = Heat transfer, W
T_m = Log mean temperature of pipe segment, °C
T_s = Ambient soil temperature, °C

L = Pipe segment length, m
R_i = Pipe insulation outer radius, mm
R_p = Pipe wall outer radius, mm
K_{ins} = Thermal conductivity of insulation, W/m/°C
K_s = Thermal conductivity of soil, W/m/°C
S = Depth of cover (pipe burial depth) to pipe centerline, mm
D = Pipe outside diameter, mm

9.2.7 Heat Transfer: Above-Ground Pipeline

An above-ground insulated pipeline can also be used to transport a heated liquid. If the pipeline is divided into segments of length L, we can calculate the heat transfer between the liquid and the ambient air using the following equations.

In English units

$$H_a = 6.28(L)(T_m - T_a)/(\text{Parm1} + \text{Parm3}) \quad (9.18)$$

$$\text{Parm3} = 1.25/[R_i(4.8 + 0.008(T_m - T_a))] \quad (9.19)$$

$$\text{Parm1} = (1/K_{ins})\text{Log}_e(R_i/R_p) \quad (9.20)$$

where

H_a = Heat transfer, Btu/hr
T_m = Log mean temperature of pipe segment, °F
T_a = Ambient air temperature, °F
L = Pipe segment length, ft
R_i = Pipe insulation outer radius, ft
R_p = Pipe wall outer radius, ft
K_{ins} = Thermal conductivity of insulation, Btu/hr/ft/°F

In SI units, Equations (9.18), (9.19), and (9.20) become

$$H_a = 6.28(L)(T_m - T_a)/(\text{Parm1} + \text{Parm3}) \quad (9.21)$$

$$\text{Parm3} = 1.25/[R_i(4.8 + 0.008(T_m - T_a))] \quad (9.22)$$

$$\text{Parm1} = (1/K_{ins})\text{Log}_e(R_i/R_p) \quad (9.23)$$

where

H_a = Heat transfer, W
T_m = Log mean temperature of pipe segment, °C
T_a = Ambient air temperature, °C
L = Pipe segment length, m
R_i = Pipe insulation outer radius, mm
R_p = Pipe wall outer radius, mm
K_{ins} = Thermal conductivity of insulation, W/m/°C

9.2.8 Frictional Heating

The frictional pressure drop causes heating of the liquid. The heat gained by the liquid due to friction is calculated using the following equations:

$$H_w = 2545\,(HHP) \tag{9.24}$$

$$HHP = (1.7664 \times 10^{-4})(Q)(Sg)(h_f)(L_m) \tag{9.25}$$

where

H_w = Frictional heat gained, Btu/hr
HHP = Hydraulic horsepower required for pipe friction
Q = Liquid flow rate, bbl/hr
Sg = Liquid specific gravity
h_f = Frictional head loss, ft/mile
L_m = Pipe segment length, miles
In SI units, Equations (9.24) and (9.25) become

$$H_w = 1000\,(Power) \tag{9.26}$$

$$Power = (0.00272)(Q)(S_g)(h_f)(L_m) \tag{9.27}$$

where

H_w = Frictional heat gained, W
Power = Power required for pipe friction, kW
Q = Liquid flow rate, m^3/hr
Sg = Liquid specific gravity
h_f = Frictional head loss, m/km
L_m = Pipe segment length, km

9.2.9 Pipe Segment Outlet Temperature

Using the formulas developed in the preceding sections and referring to the heat balance Equation (9.6), we can now calculate the temperature of the liquid at the outlet of the pipe segment as follows:
For buried pipe:

$$T_2 = (1/wC_p)[2545\,(HHP) - H_b + (wC_p)T_1] \tag{9.28}$$

For above-ground pipe:

$$T_2 = (1/wC_p)[2545(HHP) - H_a + (wC_p)T_1] \tag{9.29}$$

where

H_b = Heat transfer for buried pipe, Btu/hr from Equation (9.12)
H_a = Heat transfer for above-ground pipe, Btu/hr from Equation (9.18)

C_p = Average specific heat of liquid in pipe segment

For simplicity, we have used the average specific heat above for the pipe segment based on C_{pi} and C_{po} discussed earlier in Equations (9.8) and (9.9).

In SI units, Equations (9.28) and (9.29) can be expressed as

For buried pipe:

$$T_2 = (1/wC_p)[1000(\text{Power}) - H_b + (wC_p)T_1] \tag{9.30}$$

For above-ground pipe:

$$T_2 = (1/wC_p)[1000(\text{Power}) - H_a + (wC_p)T_1] \tag{9.31}$$

where

H_b = Heat transfer for buried pipe, W
H_a = Heat transfer for above-ground pipe, W
Power = Frictional power defined in Equation (9.27), kW

9.2.10 Liquid Heating due to Pump Inefficiency

Since a centrifugal pump is not 100% efficient, the difference between the hydraulic horsepower and the brake horsepower represents power lost. Most of this power lost is converted to heating the liquid being pumped. The temperature rise of the liquid due to pump inefficiency may be calculated from the following equation:

$$\Delta T = (H/778\,C_P)(1/E - 1) \tag{9.32}$$

where

ΔT = Temperature rise, °F
H = Pump head, ft
C_P = Specific heat of liquid, Btu/lb/°F
E = Pump efficiency, as a decimal value less than 1.0

When considering thermal hydraulics, the above temperature rise as the liquid moves through a pump station should be included in the temperature profile calculation. For example, if the liquid temperature has dropped to 120°F at the suction side of a pump station and the temperature rise due to pump inefficiency causes a 3°F rise, the liquid temperature at the pump discharge will be 123°F.

Example Problem 9.1

Calculate the temperature rise of a liquid (specific heat = 0.45 Btu/lb/°F) due to pump inefficiency as it flows through a pump. Pump head = 2450 ft and pump efficiency = 75%.

Solution

From Equation (9.32), the temperature rise is

$$\Delta T = (2450/(778 \times 0.45))(1/0.75 - 1) = 2.33°F$$

We will now use the equations discussed in this chapter to calculate the thermal hydraulic temperature profile of a crude oil pipeline.

Example Problem 9.2

A 16 in., 0.250 in. wall thickness, 50 mile long buried pipeline transports 4000 bbl/hr of heavy crude oil that enters the pipeline at 160°F. The crude oil has a specific gravity and viscosity as follows:

Temperature (°F):	100	140
Specific gravity:	0.967	0.953
Viscosity (cSt):	2277	348

Assume a pipe burial depth of 36 in. to the top of pipe and 1.5 in. insulation thickness with a thermal conductivity (K value) of 0.02 Btu/hr/ft/°F. Also assume a uniform soil temperature of 60°F with a K value of 0.5 Btu/hr/ft/°F. Using the heat balance equation calculate the outlet temperature of the crude oil at the end of the first mile segment. Assume an average specific heat of 0.45 for the crude oil.

Solution

First calculate the heat transfer for buried pipe using Equations (9.12) through (9.14):

$$\text{Parm1} = (1/0.02)\text{Log}_e(9.5/8) = 8.5925$$

$$\text{Parm2} = (1/0.5)\text{Log}_e[2 \times 44/16 + ((2 \times 44/16)^2 - 1)^{1/2}] = 4.7791$$

$$H_b = 6.28(5280)(T_m - 60)/(8.5925 + 4.7791)$$

or

$$H_b = 2479.76(T_m - 60) \quad \text{Btu/hr} \tag{9.33}$$

The log mean temperature T_m of this 1 mile pipe segment has to be approximated first, since it depends on the inlet temperature, soil temperature, and the unknown liquid temperature at the outlet of the 1 mile segment.

As a first approximation, assume the outlet temperature at the end of the 1 mile segment to be $T_2 = 150$. Calculate T_m using

Equation (9.7):

$$T_m = \frac{60 + (160 - 60) - (150 - 60)}{\text{Log}_e[(160 - 60)/(150 - 60)]} \tag{9.34}$$

or

$$T_m = 154.91°F$$

Therefore, H_b from Equation (9.33) above becomes

$$H_b = 2479.76\,(154.91 - 60) = 235{,}354 \text{ Btu/hr}$$

The frictional heating component H_w will be calculated using Equations (9.24) and (9.25). The frictional head drop h_f depends on the specific gravity and viscosity at the calculated mean temperature T_m. Using the viscosity-temperature relationship from Chapter 2, we calculate the specific gravity and viscosity at 154.91°F to be 0.9478 and 200.22 cSt, respectively.

$$\text{Reynolds number,} \quad R = \frac{92.24 \times (4000 \times 24)}{15.5 \times 200.22} = 2853$$

Using the Colebrook-White equation, the friction factor is

$$f = 0.034$$

Frictional head drop h_f will be calculated from the Darcy-Weisbach equation (3.26) as follows:

$$h_f = 0.034\,(5280 \times 12/15.5)(V^2/64.4)$$

The velocity V is calculated using Equation (3.12) as follows

$$V = 0.2859\,(4000)/(15.5)^2 = 4.76 \text{ ft/s}$$

Therefore, the frictional pressure drop is

$$h_f = 0.034\,(5280 \times 12/15.5)(4.76 \times 4.76/64.4) = 48.89 \text{ ft}$$

From Equation (9.25) the frictional horsepower (HP) is

$$HHP = (1.7664 \times 10^{-4}) \times 4000 \times 0.9478 \times 48.89 \times 1.0 = 32.74$$

Therefore frictional heating from Equation (9.24) is

$$H_w = 2545 \times 32.74 = 83{,}323 \text{ Btu/hr}$$

The mass flow rate is

$$w = 4000 \times 5.6146 \times 0.9478 \times 62.4 = 1.328 \times 10^6 \text{ lb/hr}$$

From Equation (9.30) the liquid temperature at the outlet of the 1 mile segment is

$$T_2 = (1/(1.328 \times 10^6 \times 0.45))[83{,}323 - 235{,}354 + 1.328 \times 10^6 \times 0.45 \times 160]$$

$$= [-0.255 + 160] = 159.75°F$$

This value of T_2 is used as a second approximation in Equation (9.34) to calculate a new value of T_m and subsequently the next approximation for T_2. Calculations are repeated until successive values of T_2 are within close agreement. This is left as an exercise for the reader.

It can be seen from the foregoing that manual calculation of temperatures and pressures along a heated oil pipeline is definitely a laborious process, but that can be eased using programmable calculators and personal computers.

Thermal hydraulics is very complex and calculations require utilization of some type of computer program to generate quick results. Such a program can subdivide the pipeline into short segments and calculate the temperatures, liquid properties, and pressure drops as we have seen in the examples in this chapter. Several software packages are commercially available to perform thermal hydraulics. One such package is LIQTHERM developed by SYSTEK Technologies, Inc. (www.systek.us). For a sample output report from a liquid pipeline thermal hydraulics analysis using LIQTHERM software, refer to Table A.15 in Appendix A.

9.3 Summary

We have explored the thermal effects of pipeline hydraulics in this chapter. To transport viscous liquids, they have to be heated to a temperature sometimes much higher than the ambient conditions. This temperature differential between the pumped liquid and the surrounding soil (buried pipeline) or ambient air (above-ground pipeline) causes heat transfer to occur, resulting in temperature variation of the liquid along the pipeline. Unlike isothermal flow, where the liquid temperature is uniform throughout the pipeline, heated pipeline hydraulics requires subdividing the pipeline into short segments and calculating pressure drops based on liquid properties at the average temperature of each segment. We illustrated this using an example that showed how the temperature varies along the pipeline.

The concept of LMTD was introduced for determining a more accurate average segment temperature. Also, a method to compute the heat transfer between the liquid and the surrounding medium was shown taking into account thermal conductivities of pipe, soil, and insulation, if present. The heating of liquid due to friction was also quantified, as was the liquid heating associated with pump inefficiency.

It was pointed out that unlike isothermal hydraulics, thermal hydraulics is a complex phenomenon that requires computer methods to correctly solve equations for temperature variation and pressure drop calculations.

9.4 Problems

9.4.1 An 8 in. nominal diameter pipe is used to move heavy crude oil from a heated storage tank to a refinery 12 miles away. The inlet temperature is 180°F and the crude oil has the following characteristics:

Temperature (°F):	100	180
Specific gravity:	0.985	0.912
Viscosity (cSt):	4000	25

If the outlet temperature cannot drop below 150°F, what minimum flow rate needs to be maintained in the pipeline? Use the MIT equation and 0.002 in. pipe roughness, 70°F soil temperature, 36 in. burial depth, and 0.25 in. insulation, with K = 0.02 Btu/hr/ft/°F. The K value for soil = 0.6 Btu/hr/ft/°F. Pipeline pressures are limited to ANSI 600 rating (1440 psi).

9.4.2 In Problem 9.4.1, if insulation were absent what minimum flow rate could be tolerated? Compare the HP requirements in both cases. Assume 15 psi pump suction pressure.

9.4.3 From Problem 9.4.1 develop a set of data points for flow rate versus pressure required for plotting a system head curve. Assume 50 psi delivery pressure.

10

Flow Measurement

In this chapter we discuss the various methods and instruments used in the measurement of liquid that flows through a pipeline. The formulas used for calculating the liquid velocities, flow rates, etc., from the pressure readings, their limitations, and the degree of accuracy attainable will be covered for some of the more commonly used instruments. Considerable work is currently being done in this field of flow measurement to improve the accuracy of instruments, particularly when custody transfer of products is involved. For more detailed analysis of the various flow measurement devices, the reader is referred to the publications listed in the References section.

10.1 History

Measurement of flow of liquids has been going on for centuries. The Romans used some form of flow measurement to allocate water from the aqueducts to the houses in their cities. This was necessary to control the quantity of water used by the citizens and prevent waste. Similarly, it is reported that the Chinese used to measure salt water to the brine pots that were used for salt production. In later years, a commodity had to be measured so that it could be properly allocated and the ultimate user charged for it appropriately. Today, gasoline is dispensed from meters at gas stations and the recipient is billed according to the volume of gasoline so measured. Water companies measure water consumed by a household using water meters, while natural gas for residential and industrial consumers is measured by gas meters. In all these instances, the objective is to ensure that the supplier gets paid for the commodity and the user of the commodity is billed for the product at the agreed price. In addition, in industrial processes that involve the use of liquids and gases to perform a specific function, accurate quantities need to be dispensed so that the desired effect of the processes may be realized. In most cases involving consumers, certain regulatory or public agencies (for example, the department of weights and measures) periodically check flow measurement devices to ensure that they are performing accurately and if necessary calibrate them against a very accurate master device.

10.2 Flow Meters

Several types of instruments are available to measure the flow rate of a liquid in a pipeline. Some measure the velocity of flow, while others directly measure the volume flow rate or the mass flow rate.

The following flow meters are used in the pipeline industry:

Venturi meter
Flow nozzle
Orifice meter
Flow tube
Rotameter
Turbine meter
Vortex flow meter
Magnetic flow meter
Ultrasonic flow meter
Positive displacement meter
Mass flow meter
Pitot tube

The first four items listed above are called variable-head meters since the flow rate measured is based on the pressure drop due to a restriction in the meter, which varies with the flow rate. The last item, the Pitot tube, is also called a velocity probe, since it actually measures the velocity of the liquid. From the measured velocity, we can calculate the flow rate using the conservation of mass equation, discussed in an earlier chapter.

$$\text{Mass flow} = (\text{Flow rate}) \times (\text{density})$$
$$= (\text{Area}) \times (\text{Velocity}) \times (\text{density})$$

or

$$M = Q_\rho$$
$$= AV_\rho$$

Since the liquid density is practically constant, we can state that:

$$Q = AV \qquad (10.2)$$

where

A = Cross-sectional area of flow
V = Velocity of flow
ρ = Liquid density

We will discuss the principle of operation and the formulas used for some of the more common meters described above. For a more detailed discussion on flow meters and flow measurement, the reader is referred to one of the fine texts listed in the References section.

10.3 Venturi Meter

The Venturi meter, also known as a Venturi tube, belongs to the category of variable-head flow meters. The principle of a Venturi meter is depicted in Figure 10.1. This type of a Venturi meter is also known as the Herschel type and consists of a smooth gradual contraction from the main pipe size to the throat section, followed by a smooth, gradual enlargement from the throat section to the original pipe diameter.

The included angle from the main pipe to the throat section in the gradual contraction is generally in the range of $21° \pm 2°$. Similarly, the gradual expansion from the throat to the main pipe section is limited to a range of 5° to 15° in this design of Venturi meter. This construction results in minimum energy loss, causing the discharge coefficient (discussed later) to approach the value of 1.0. This type of a Venturi meter is generally rough cast with a pipe diameter range of 4.0 in. to 48.0 in. The throat diameter may vary considerably, but the ratio of the throat diameter to the main pipe diameter (d/D), also known as the beta ratio, represented by the symbol β, should range between 0.30 and 0.75.

The Venturi meter consists of a main piece of pipe which decreases in size to a section called the throat, followed by a gradually increasing size back to the original pipe size. The liquid pressure at the main pipe section 1

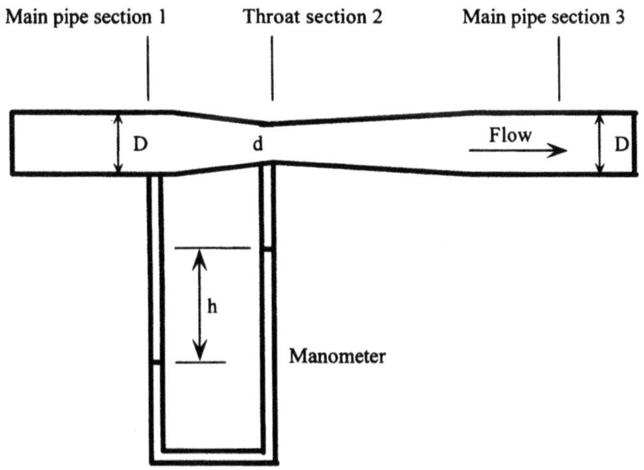

Figure 10.1 Venturi meter.

is denoted by P_1 and that at the throat section 2 is represented by P_2. As the liquid flows through the narrow throat section, it accelerates to compensate for the reduction in area, since the volume flow rate is constant ($Q = AV$ from Equation 10.2). As the velocity increases in the throat section, the pressure decreases (Bernoulli's equation). As the flow continues past the throat, the gradual increase in the flow area results in a reduction in flow velocity back to the level at section 1; correspondingly the liquid pressure will increase to some value higher than at the throat section.

We can calculate the flow rate using Bernoulli's equation, introduced in Chapter 2, along with the continuity equation, based on the conservation of mass as shown in Equation (10.1). If we consider the main pipe section 1 and the throat section 2 as reference, and apply Bernoulli's equation for these two sections, we get

$$P_1/\gamma + V_1^2/(2g) + Z_1 = P_2/\gamma + V_2^2/(2g) + Z_2 + h_L \tag{10.3}$$

And from the continuity equation

$$Q = A_1V_1 = A_2V_2 \tag{10.4}$$

where γ is the specific weight of liquid and h_L represents the pressure drop due to friction between sections 1 and 2.

Simplifying the above equations yields the following:

$$V_1 = \sqrt{}[2g(P_1 - P_2)/\gamma + (Z_1 - Z_2) - h_L]/[(A_1/A_2)^2 - 1] \tag{10.5}$$

The elevation difference $Z_1 - Z_2$ is negligible even if the Venturi meter is positioned vertically. We will also drop the friction loss term h_L and include it in a coefficient C, called the discharge coefficient, and rewrite Equation (10.5) as follows:

$$V_1 = C\sqrt{[2g(P_1 - P_2)/\gamma]/[(A_1/A_2)^2 - 1]} \tag{10.6}$$

The above equation gives us the velocity of the liquid in the main pipe section 1. Similarly, the velocity in the throat section, V_2, can be calculated using Equation (10.4) as follows:

$$V_2 = C\sqrt{[2g(P_1 - P_2)/\gamma]/[1 - (A_2/A_1)^2]} \tag{10.7}$$

The volume flow rate Q can now be calculated using Equation (10.4) as follows:

$$Q = A_1 V_1$$

Therefore,

$$Q = CA_1\sqrt{[2g(P_1 - P_2)/\gamma]/[(A_1/A_2)^2 - 1]} \tag{10.8}$$

Since the beta ratio $\beta = d/D$ and $A_1/A_2 = (D/d)^2$ we can write the above equations in terms of the beta ratio as follows:

$$Q = CA_1\sqrt{[2g(P_1 - P_2)/\gamma]/[(1/\beta)^4 - 1]} \tag{10.9}$$

It can be seen by examining Equations (10.5), (10.6) and (10.7) that the discharge coefficient C actually represents the ratio of the velocity of liquid through the Venturi meter to the ideal velocity when the energy loss (h_L) is zero. C must therefore be a number less than 1.0. The value of C depends on the Reynolds number in the main pipe section 1 and is shown graphically in Figure 10.2. For a Reynolds number greater than 2×10^5 the value of C remains constant at 0.984.

For smaller piping (2 to 10 in.), Venturi meters are machined and hence have a better surface finish than the larger, rough cast meters. These smaller Venturi meters have a C value of 0.995 for Reynolds number greater than 2×10^5.

10.4 Flow Nozzle

A typical flow nozzle is illustrated in Figure 10.3. It consists of the main pipe section 1, followed by a gradual reduction in flow area and a subsequent short cylindrical section, and finally an expansion to the main pipe section 3.

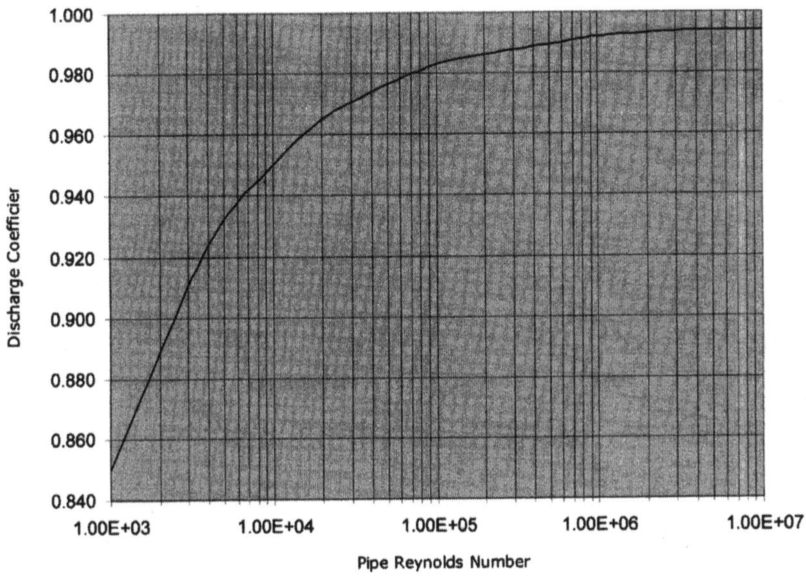

Figure 10.2 Venturi meter discharge coefficient.

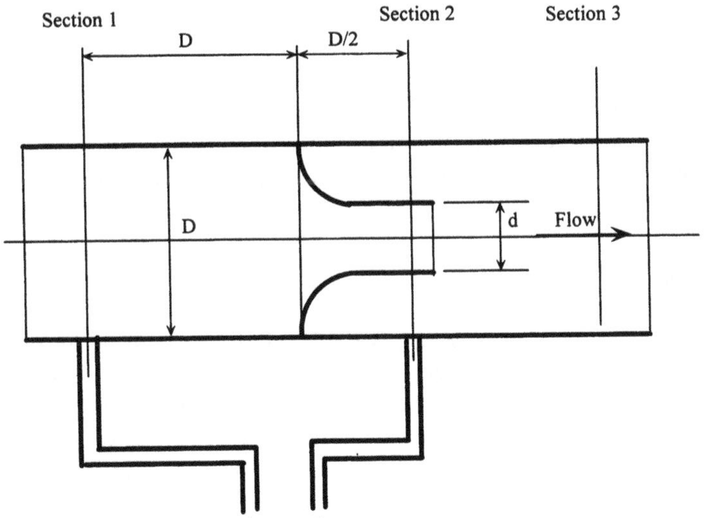

Figure 10.3 Flow nozzle. (From Mott, Robert L., *Applied Fluid Mechanics, 5th ed.*, Copyright 2000. Reprinted by permission of Pearson Education, Inc., Upper Saddle River, NJ.)

The American Society of Mechanical Engineers (ASME) and the International Standards Organization (ISO) have defined the geometries of these flow nozzles and published equations to be used with them. Due to the smooth gradual contraction from the main pipe diameter D to the nozzle diameter d, the energy loss between sections 1 and 2 is very small. We can apply the same Equations (10.6) through (10.8) for the Venturi meter to the flow nozzle also. The discharge coefficient C for the flow nozzle is found to be 0.99 or better for Reynolds numbers above 10^6. At lower Reynolds numbers there is greater energy loss immediately following the nozzle throat, due to sudden expansion, and hence C values are lower.

Depending on the beta ratio and the Reynolds number, the discharge coefficient C can be calculated from the equation

$$C = 0.9975 - 6.53\sqrt{(\beta/R)} \qquad (10.10)$$

where

$$\beta = d/D$$

and R is the Reynolds number based on the pipe diameter D.

Compared with the Venturi meter, the flow nozzle is more compact, since it does not require the length for a gradual decrease in diameter at the throat, or the additional length for the smooth, gradual expansion from the throat to the main pipe size. However, there is more energy loss (and, therefore, pressure head loss) in the flow nozzle, due to the sudden expansion from the nozzle diameter to the main pipe diameter. The latter causes greater turbulence and eddies compared with the gradual expansion in the Venturi meter.

10.5 Orifice Meter

An orifice meter consists of a flat plate that has a sharp-edged hole accurately machined in it and placed concentrically in a pipe, as shown in Figure 10.4. As liquid flows through the pipe, the flow suddenly contracts as it approaches the orifice and then suddenly expands after the orifice back to the full pipe diameter. This forms a vena contracta or a throat immediately past the orifice. This reduction in flow pattern at the vena contracta causes increased velocity and hence lower pressure at the throat, similar to the Venturi meter discussed earlier.

The pressure difference between section 1, with the full flow, and section 2 at the throat can then be used to measure the liquid flow rate, using equations developed earlier for the Venturi meter and the flow nozzle. Due to the sudden contraction at the orifice and the subsequent sudden

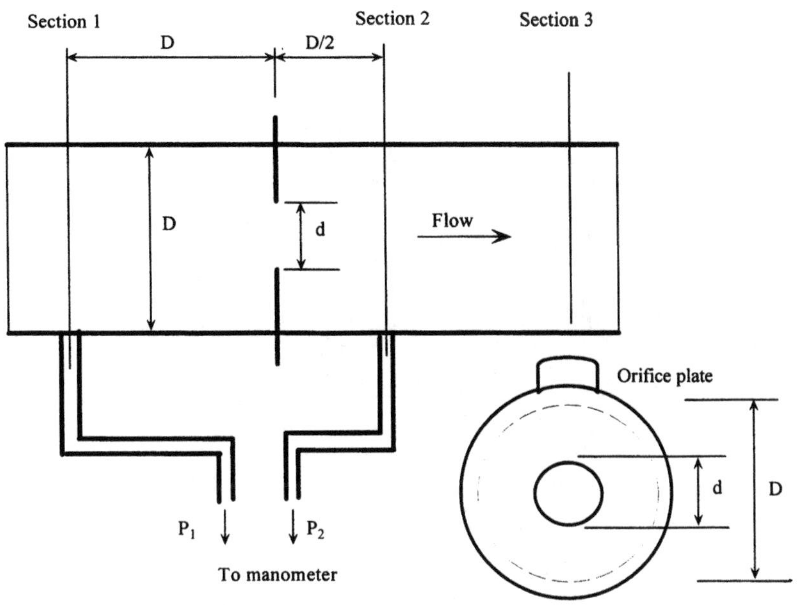

Figure 10.4 Orifice meter. (From Mott, Robert L., *Applied Fluid Mechanics, 5th ed.*, Copyright 2000. Reprinted by permission of Pearson Education, Inc., Upper Saddle River, NJ.)

Table 10.1 Pressure Taps for Orifice Meter

	Inlet pressure tap, P_1	Outlet pressure tap, P_2
1	One pipe diameter upstream from plate	One-half pipe diameter downstream of inlet face of plate
2	One pipe diameter upstream from plate	At vena contracta
3	Flange taps, 1 in. upstream from plate	Flange taps, 1 in. downstream from outlet face of plate

expansion after the orifice, the coefficient of discharge C for the orifice meter is much lower than that of a Venturi meter or a flow nozzle. In addition, depending on the pressure tap locations in section 1 and section 2, the value of C is different for orifice meters.

There are three possible pressure tap locations for an orifice meter, as listed in Table 10.1. Figure 10.5 shows the variation of C with the beta ratio d/D for various values of pipe Reynolds number.

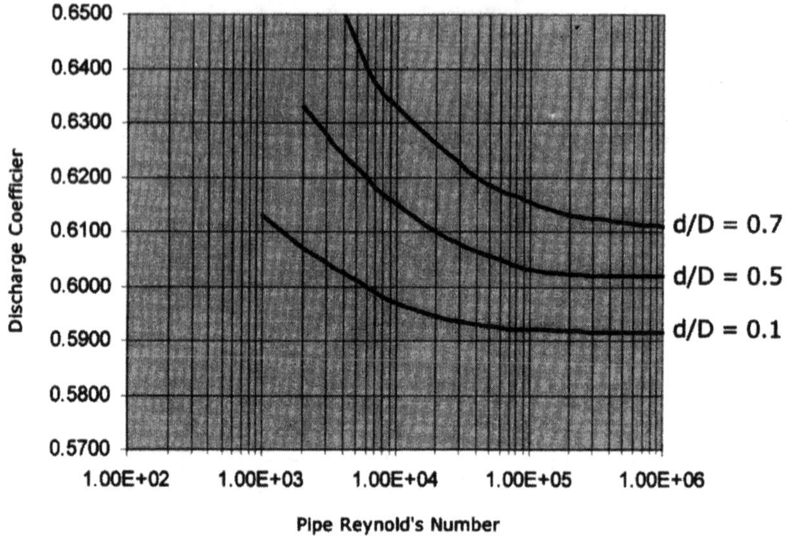

Figure 10.5 Orifice meter discharge coefficient. (From Mott, Robert L., *Applied Fluid Mechanics, 5th ed.*, Copyright 2000. Reprinted by permission of Pearson Education, Inc., Upper Saddle River, NJ.)

Comparing the three types of flow meter discussed above, we can conclude that the orifice meter has the highest energy loss due to the sudden contraction followed by the sudden expansion. On the other hand, the Venturi meter has a lower energy loss compared with the flow nozzle due to the smooth, gradual reduction at the throat followed by the smooth gradual expansion after the throat.

Flow tubes are proprietary, variable-head flow meters that are streamlined in design to cause the least amount of energy loss. They are manufactured by different companies with their own proprietary designs.

10.6 Turbine Meters

Turbine meters are used in a wide variety of applications. The food industry uses turbine meters for measurement of milk, cheese, cream, syrups, vegetable oils, etc. Turbine meters are also used in the oil industry.

Basically the turbine meter is a velocity-measuring instrument. Liquid flows through a free-turning rotor that is mounted co-axially inside the meter. Upstream and downstream of the meter, certain lengths of straight piping are required to ensure smooth velocities through the meter. The liquid striking the rotor causes it to rotate, the velocity of rotation being proportional to the flow rate. From the rotor speed measured and the flow area, we can compute the flow rate through the meter. The turbine meter must be calibrated since flow depends on the friction, turbulence, and the manufacturing tolerance of the rotor parts.

For liquids with viscosity close to that of water the range of flow rates is 10:1. With higher- and lower-viscosity liquids it drops to 3:1. Density effect is similar to viscosity.

A turbine meter requires a straight length of pipe before and after the meter. The straightening vanes located in the straight lengths of pipe helps eliminate swirl in flow. The upstream length of straight pipe has to be 10D in length, where D is the pipe diameter. After the meter, the straight length of pipe is 5D long. Turbine meters may be of a two-section type or three-section type, depending upon the number of pieces the meter assembly is composed of. For liquid flow measurement, the API *Manual of Petroleum Measurement* standard must be followed. In addition, bypass piping must also be installed to isolate the meter for maintenance and repair. To maintain flow measurements on a continuous basis for custody transfer, an entire spare meter unit will be required on the bypass piping, to be used when the main meter is taken out of service for testing and maintenance.

From the turbine meter reading at flowing temperature, the flow rate at some base temperature (such as 60°F) is calculated using the following equation:

$$Q_b = Q_f \times M_f \times F_t \times F_p \tag{10.11}$$

where

Q_b = Flow rate at base conditions, such as 60°F and 14.7 psi

Q_f = Measured flow rate at operating conditions, such as 80°F and 350 psi

M_f = Meter factor for correcting meter reading, based on meter calibration data

F_t = Temperature correction factor for correcting from flowing temperature to the base temperature

F_p = Pressure correction factor for correcting from flowing pressure to the base pressure

10.7 Positive Displacement Meter

The positive displacement (PD) meter is most commonly used in residential applications for measuring water and natural gas consumption. PD meters can measure small flows with good accuracy. These meters are used when consistently high accuracy is required under steady flow through a pipeline. PD meters are accurate within ±1% over a flow range of 20:1. They are suitable for batch operations, for mixing and blending of liquids that are clean without deposits, and for use with noncorrosive liquids.

PD meters basically operate by measuring fixed quantities of the liquid in containers that are alternately filled and emptied. The individual volumes are then totaled and displayed. As the liquid flows through the meter, the force from the flowing stream causes a pressure drop in the meter, which is a function of the internal geometry. The PD meter does not require the straightening vanes as with a turbine meter and hence is a more compact design. However PD meters can be large and heavy compared with an equivalent turbine meter. Also, jamming can occur when the meter is stopped and restarted. This necessitates some form of bypass piping with valves to prevent a pressure rise and damage to meter. The accuracy of a PD meter depends on clearance between the moving and stationary components. Hence precision machined parts are required to maintain accuracy. PD meters are not suitable for slurries or liquids with suspended particles that could jam the components and cause damage to the meter. There are several types of PD meter, including the reciprocating piston type, rotating disk, rotary piston, sliding vane, and rotary vane type.

Many modern petroleum installations employ PD meters for crude oil and refined products. These have to be calibrated periodically, to ensure accurate flow measurement. Most PD meters used in the oil industry are tested at regular intervals, such as daily or weekly, using a master meter called a meter prover system that is installed as a fixed unit along with the PD meter.

Example Problem 10.1

A Venturi meter is used for measuring the flow rate of water at 70°F. The flow enters a 6.625 in. pipe, 0.250 in. wall thickness; the throat diameter is 2.4 in. The Venturi meter is of the Herschel type and is rough cast with a mercury manometer. The manometer reading shows a pressure difference of 12.2 in. of mercury. Calculate the flow velocity in the pipe and the volume flow rate. Use a specific gravity of 13.54 for mercury.

Solution

At 70°F the specific weight and viscosity are:

$$\gamma = 62.3 \text{ lb/ft}^3$$

and

$$\nu = 1.05 \times 10^{-5}\ \text{ft}^2/\text{s}$$

Assume first that the Reynolds number is greater than 2×10^5. Therefore, the discharge coefficient C will be 0.984 from Figure 10.2. The beta ratio $\beta = 2.4/(6.625 - 0.5) = 0.3918$ and therefore β is between 0.3 and 0.75.

$$A_1/A_2 = (1/0.3918)^2 = 6.5144$$

The manometer reading will give us the pressure difference, by equating the pressures at the two points in the manometer:

$$P_1 + \gamma_w(y + h) = P_2 + \gamma_w y + \gamma_m h$$

where y is depth of the higher mercury level below the centerline of the Venturi meter and γ_w and γ_m are the specific weights of water and mercury respectively.
Simplifying the above, we get:

$$(P_1 - P_2)/\gamma w = (\gamma_m/\gamma_w - 1)h$$
$$= (13.54 \times 62.4/62.3 - 1) \times 12.2/12 = 12.77\ \text{ft}$$

And using Equation (10.6), we get

$$V_1 = 0.984\sqrt{[(64.4 \times 12.77)/(6.5144^2 - 1)]}$$
$$= 4.38\ \text{ft/s}$$

Now we check the value of the Reynolds number:

$$R = 4.38 \times (6.125/12)/1.05 \times 10^{-5}$$
$$= 2.13 \times 10^5$$

Since R is greater than 2×10^5 our assumption for C is correct. The volume flow rate is

$$Q = A_1 V_1$$

$$\text{Flow rate} = 0.7854(6.125/12)^2 \times 4.38 = 0.8962\ \text{ft}^3/\text{s}$$

or

$$\text{Flow rate} = 0.8962 \times (1728/231) \times 60 = 402.25\ \text{gal/min}$$

Example Problem 10.2

An orifice meter is used to measure the flow rate of kerosene in a 2 in. schedule 40 pipe at 70°F. The orifice is 1.0 in. diameter. Calculate the volume flow rate if the pressure difference across the orifice is 0.54 psi. Specific gravity of kerosene = 0.815 and viscosity = 2.14×10^{-5} ft^2/s.

Solution

Since the flow rate depends on the discharge coefficient C, which in turn depends on the beta ratio and the Reynolds number, we will have to solve this problem by trial and error.

The 2 in. schedule 40 pipe is 2.375 in. outside diameter and 0.154 in. wall thickness. The beta ratio

$$\beta = 1.0/(2.375 - 2 \times 0.154) = 0.4838$$

First, we assume a value for the discharge coefficient C (say 0.61) and calculate the flow rate from Equation (10.6). We then calculate the Reynolds number and obtain a more correct value of C from Figure 10.5. Repeating the method a couple of times will yield a more accurate value of C and finally the flow rate.

The density of kerosene = $0.815 \times 62.3 = 50.77$ lb/ft^3

Ratio of areas

$$A_1/A_2 = (1/0.4838)^2 = 4.2717$$

Using Equation (10.6), we get

$$V_1 = 0.61\sqrt{[(64.4 \times 0.54 \times 144/(50.77)/(4.2717^2 - 1)]} = 1.4588 \text{ ft/s}$$

$$\text{Reynolds number} = 1.4588(2.067/12)/(2.14 \times 10^{-5})$$
$$= 11{,}742$$

Using this value of the Reynolds number, we obtain from Figure 10.5 the discharge coefficient C = 0.612. Since we earlier assumed C = 0.61 we were not too far off.

Recalculating based on the new value of C:

$$V_1 = 0.612\sqrt{[(64.4 \times 0.54 \times 144/(50.77)/(4.2717^2 - 1)]} = 1.46 \text{ ft/s}$$

$$\text{Reynolds number} = 1.46(2.067/12)/(2.14 \times 10^{-5})$$
$$= 11{,}751$$

This is quite close to the previous value of the Reynolds number. Therefore we will use the last calculated value for velocity. The volume flow rate is

$$Q = A_1 V_1$$

$$\text{Flow rate} = 0.7854(2.067/12)^2 \times 1.46 = 0.034\ \text{ft}^3/\text{s}$$

or

$$\text{Flow rate} = 0.034 \times (1728/231) \times 60 = 15.27\ \text{gal/min}$$

10.8 Summary

In this chapter we discussed the more commonly used devices for the measurement of liquid pipeline flow rates. Variable-head flow meters, such as the Venturi meter, flow nozzle and orifice meter, and the equations for calculating the velocities and flow rate from the pressure drop were explained. The importance of the discharge coefficient and how it varies with the Reynolds number and the beta ratio were also discussed. A trial-and-error method for calculating the flow rate through an orifice meter was illustrated using an example. For a more detailed description and analysis of flow meters, the reader should refer to any of the standard texts used in the industry. Some of these are listed in the References section.

10.9 Problems

10.9.1 Water flow through a 4 in. internal diameter pipeline is measured using a Venturi tube. The flow rate expected is 300 gal/min. Specific weight of water is 62.4 lb/ft^3 and viscosity is 1.2×10^{-5} ft^2/s.

(a) Calculate the upstream Reynolds number.
(b) If the pressure differential across the meter is 10 psi, determine the dimensions of the meter.
(c) Calculate the throat Reynolds number.

10.9.2 A 12 in. Venturi meter with a 7 in. throat is used for measuring flow of diesel at 60°F. The differential pressure reads 70 in. of water. Calculate the volume and mass flow rates. Specific gravity of diesel is 0.85 and viscosity is 5 cSt.

10.9.3 An orifice meter with a beta ratio of 0.70 is used for measuring crude oil (0.895 specific gravity and 15 cSt viscosity). The pipeline is 10.75 in. diameter, 0.250 in. wall thickness and the line pressure is 250 psig. With pipe taps the differential pressure is 200 in. of water. Calculate the flow rate of crude oil.

10.9.4 Water flows through a Venturi tube of 300 mm main pipe diameter and 150 mm throat diameter at 150 m^3/hr. The differential manometer deflection is 106 cm. The specific gravity of the manometer liquid is 1.3. Calculate the discharge coefficient of the meter.

11

Unsteady Flow in Pipelines

Throughout the last 10 chapters we have addressed steady flow of liquids in pipelines. This means that the pipeline flow rates, velocities, and pressures at any point along the pipeline do not change with time. In other words, flow is not time-dependent. In reality this may not be true. Many factors affect pipeline operation. Even though volumes may be taken out of storage tanks at a constant rate and pump stations pump the liquid at a uniform rate there will always be some environmental or mechanical factors that could result in changes to the steady-state flow in the pipeline. In this chapter we briefly discuss unsteady flow and pipeline transients. A detailed analysis of unsteady flow and pipeline transient pressures is beyond the scope of this book. The reader is advised to refer to more specialized texts on unsteady flow, listed in the References section.

11.1 Steady Versus Unsteady Flow

In steady-state flow we assume the hydraulic gradient, representing the variation in pipeline pressures along the pipeline, to be fixed as long as the same liquid is being pumped at the same flow rate. Also, in steady-state flow we assume that liquids are incompressible. In reality, liquids *are* compressible, to some extent. The compressibility effect can cause transients or surges

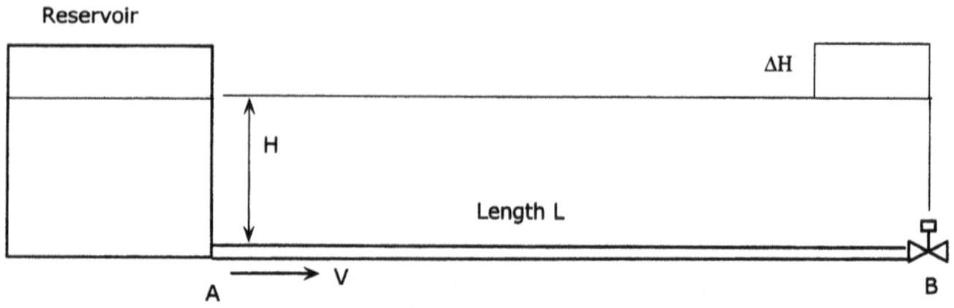

Figure 11.1 Transient with sudden valve closure.

in otherwise steady-state flow. If, during the course of operation, different liquids enter the pipeline at different times, such as in a batched pipeline, the pressures, velocities, and flow rates will vary with time, resulting in changes to the hydraulic pressure gradient.

Unsteady flow occurs when the flow rate, velocity, and pressure at a point in a pipeline change with time. Examples of such pressure transients include the effect of opening and closing valves, starting and stopping a pump, or starting and stopping pipeline injections or deliveries. A basic example of unsteady flow would be a situation when a downstream valve on a pipeline is suddenly closed, as shown in Figure 11.1.

11.2 Transient Flow due to Valve Closure

To illustrate the effect of unsteady flow and transient pressure, we will consider a simple pipeline system from a storage tank at A, to a valve at the end of the pipe at B, as shown in Figure 11.1. The valve is located at a distance L from the tank. The pipeline is assumed to be horizontal, with no elevation changes and uniform diameter D. In order to simplify the problem, we will further assume that the friction in the pipe is negligible. Later we will consider friction and its effect on transient flow. Since friction is ignored, the steady-state hydraulic gradient line is horizontal.

Let us assume that initially, at time $t = 0$, the valve at B is fully open, steady-state flow Q exists from A to B, and the liquid velocity throughout the pipe is V. The tank head is assumed to be H. The velocity is assumed to be positive in the downstream direction from A to B. If nothing changed and the valve at B were left open long enough, the tank head H would be dissipated by flow through the valve.

When we suddenly close the valve at B, we introduce a transient or surge into the pipeline. As the valve closes, pressure, velocity, and flow variations occur both upstream and downstream of the valve. Since we are only interested in the pipe section from A to B, we will ignore the effect downstream of the valve at B. Closing the valve suddenly causes the velocity of flow at B to become zero instantaneously. This sudden drop in velocity at the valve causes the pressure head at B to increase suddenly by some value ΔH. This pressure rise is equal to the change in momentum of the liquid from a velocity of V to zero.

If the velocity changes from V to zero,

$$\Delta V = 0 - V = -V$$

The negative sign indicates a decrease in velocity. The magnitude of ΔH is given by

$$\Delta H = aV/g \tag{11.1}$$

where

a = Velocity of propagation of pressure wave, ft/s
V = Velocity of liquid flow, ft/s
g = Acceleration due to gravity, ft/s^2

If instead of closing the valve fully, we close it such that the velocity drops from V to V_1

$$\Delta V = V_1 - V$$

The corresponding head rise at the valve is

$$\Delta H = a(V_1 - V)/g \tag{11.2}$$

The pressure head increase ΔH at B causes a slight expansion of the pipe and also an increase in liquid density, which depends upon the bulk modulus of the liquid, pipe size, and pipe material. For most pipelines, this pipe expansion and increase in liquid density (decrease in liquid volume) will be insignificant. Depending on the properties of the liquid and pipe, the pressure increase propagates upstream from the valve at the speed of sound (wave speed), a, generally in the range of 2000 to 5000 ft/s. The wave traveling at the speed of sound reaches the tank in time L/a. At this point in time the velocity of liquid throughout the pipe has become zero and, correspondingly, the pressure everywhere has reached $H + \Delta H$. The consequence of this is that the liquid has compressed somewhat and the pipe has stretched as well.

The above scenario is an unstable condition since the tank head is H but the pipeline head is higher ($H + \Delta H$). Due to this pressure differential, the liquid starts to flow from the pipeline into the tank with velocity $-V$.

The velocity is thus changed from zero to $-V$ which causes the pressure to drop from $(H + \Delta H)$ to H. Therefore, a negative pressure wave travels toward the valve B. Since friction is negligible, the magnitude of this reverse flow velocity is exactly equal to the original velocity V.

At time $t = 2L/a$, the pressure in the pipeline returns to the steady-state value H, but reverse flow continues. However, since the valve is closed, no flow can take place upstream of the valve. Consequently, the pressure head drops by the amount ΔH, forcing the reverse flow velocity to zero. This causes the pipe to shrink and the liquid expands.

At time $t = 3L/a$, this transient has reached the tank and the flow velocity is now zero throughout the pipe. The pipe pressure, however, has dropped to a value ΔH below that of the tank. This causes flow from A to B equal to the original steady-state flow. During this process the pipe pressure returns to the original steady-state value.

Finally, at time $t = 4L/a$, the pressure wave from A has reached the valve at B and the flow velocity reaches the original steady-state value V. The total time elapsed $(4L/a)$ is defined as one wave cycle, or the theoretical period of the pipeline. As the valve is completely closed, the preceding sequence of events starts again at $t = 4L/a$. Since we assumed a frictionless system, the process continues indefinitely and the conditions are repeated at an interval of $4L/a$. However, in reality, due to pipe friction the transient pressure waves are dissipated over a definite period of time and the liquid becomes stationary with no flow anywhere and a pressure head equal to that of the tank.

Figure 11.2 shows a plot of the pressures at the valve at various times, beginning with the valve closure at time $t = 0$ to time $t = 12L/a$ (three wave periods). When friction losses are taken into consideration, the variation of pressure at the valve with time will be as shown in Figure 11.3. It can be seen from Figure 11.3 that, due to pipe friction, the transient pressure waves will die down over a definite period of time and the liquid will reach equilibrium with zero flow throughout the pipeline with the pressure head exactly equal to the initial tank head.

As mentioned before, examples of unsteady flow include filling a pipeline, power failure and pump shut-down, and opening and closing of valves. In summary we can say that unsteady flow conditions occur when the pipeline flow rate continually varies with time or variations in flow between two steady-state conditions occur. Unsteady flow conditions may be referred to as transient flow. Any change from a regular steady-state flow condition can cause transients. Slow transients are called surges. Depending on the magnitude of the transient and the rate at which the transient occurs, pipeline pressures may exceed steady-state pressures and sometimes violate the maximum allowable operating pressure (MAOP) in a pipeline.

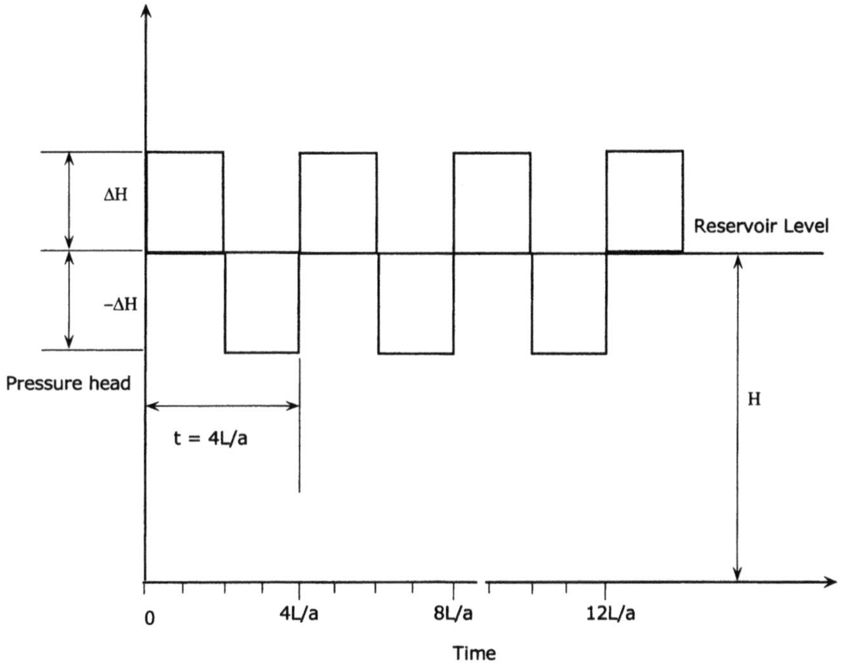

Figure 11.2 Pressure variation at valve: friction neglected.

Design codes and federal regulations dictate that the pressure surge cannot exceed 110% of the MAOP. Therefore, for a pipeline with an MAOP of 1400 psi, during transient situations the highest pressure reached anywhere in the pipeline cannot exceed $1400 + 140 = 1540$ psi.

To control surge pressures and avoid pipeline overpressure, relief valves may be used that are set to relieve the pipe pressure by allowing a certain amount of flow to a relief tank.

11.3 Wave Speed in Pipeline

We have seen from Equation (11.1) that the instantaneous head rise due to sudden stoppage of flow is

$$\Delta H = aV/g$$

The above equation can be derived easily considering a control element of flow in the tank-pipe example discussed in Section 11.2. However, we will not attempt to derive the equation here. The reader is referred instead to one of the fine books on unsteady flow listed in the References section. At this

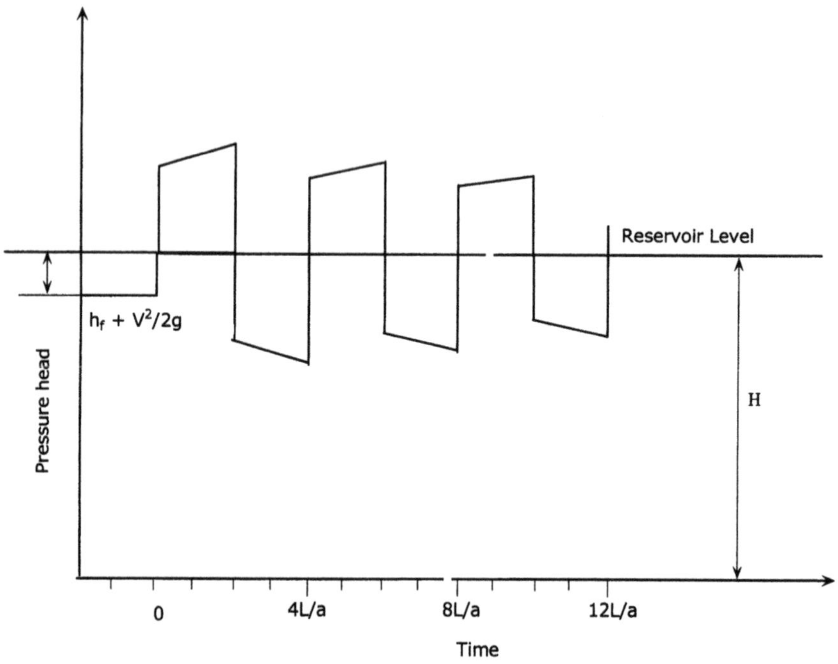

Figure 11.3 Pressure variation at valve: friction considered.

point we will assume that the above equation is correct and proceed to apply it to some practical situations.

Since the transient head rise ΔH is directly dependent on the wave speed a from Equation (11.1), we must accurately determine its value for each pipeline.

It has been found that the wave speed in a liquid depends on the density and bulk modulus of the liquid. It depends also on the elasticity of the pipe material, pipe diameter, and wall thickness. Also, if the liquid contains free air or gas, the wave speed will be affected by these. An approximate value of the wave speed is calculated from

$$a = (K/\rho)^{1/2}$$

where

a = Wave speed, ft/s
K = Bulk modulus of liquid, psi
ρ = Density of liquid, slugs/ft^3

However, the above value of wave speed does not take into account the elastic nature of the pipe. Hence in the following discussion the equation will be modified to account for the pipe material and its elastic properties.

Consider an extreme case of a pipeline system where the pipe is considered rigid and the liquid incompressible. Incompressible liquid means infinite wave speed, which is impossible. Since no pipe is totally rigid, the infinite wave speed situation is not possible.

According to Wylie and Streeter (see References), a general equation for calculating the wave speed in terms of liquid properties, pipe size, and elasticity of pipe material is as follows:

$$a = \frac{(K/\rho)^{1/2}}{[1 + C(K/E)(D/t)]^{1/2}} \tag{11.3}$$

where

a = Wave speed, ft/s
K = Bulk modulus of liquid, psi
ρ = Density of liquid, slugs/ft^3
C = Restraint factor, dimensionless
D = Pipe outside diameter, in.
t = Pipe wall thickness, in.
E = Young's modulus of pipe material, psi

An equivalent formula in SI units is

$$a = \frac{(K/\rho)^{1/2}}{[1 + C(K/E)(D/t)]^{1/2}} \tag{11.4}$$

where

a = Wave speed, m/s
K = Bulk modulus of liquid, kPa
ρ = Density of liquid, kg/m^3
C = Restraint factor, dimensionless
D = Pipe outside diameter, mm
t = Pipe wall thickness, mm
E = Young's modulus of pipe material, kPa

The Young's modulus is also referred to as the modulus of elasticity of the pipe material.

The restraint factor C depends on the type of pipe condition. The following three cases are possible:

Case 1: Pipe is anchored at the upstream end only
Case 2: Pipe is anchored against any axial movements
Case 3: Each pipe section is anchored with expansion joints

The calculated value of the wave speed depends on the restraint factor C which in turn depends on the type of pipe support described above. It is found that the restraint type has less than 10% effect on the magnitude of wave speed.

The restraint factor C is defined in terms of the Poisson's ratio μ of the pipe material as shown below (see Wylie and Streeter in References):

$$C = 1 - 0.5\mu \qquad \text{for case 1} \tag{11.5}$$

$$C = 1 - \mu^2 \qquad \text{for case 2} \tag{11.6}$$

$$C = 1.0 \qquad \text{for case 3} \tag{11.7}$$

where μ = Poisson's ratio for pipe material, usually in the range of 0.20 to 0.45 (for steel pipe $\mu = 0.30$).

It must be noted that the above values of the restraint factor C are applicable to thin-walled elastic pipes with a pipe diameter to thickness ratio (D/t) greater than 25. For thick-walled pipes with a D/t ratio less than 25, the following C values may be used (see Wylie and Streeter in References):

Case 1

$$C = \frac{2t}{D}(1 + \mu) + \frac{D(1 - 0.5\mu)}{D + t} \tag{11.8}$$

Case 2

$$C = \frac{2t}{D}(1 + \mu) + \frac{D(1 - \mu^2)}{D + t} \tag{11.9}$$

Case 3

$$C = \frac{2t}{D}(1 + \mu) + \frac{D}{D + t} \tag{11.10}$$

Example Problem 11.1

Water flows through a steel pipe at velocity of 6 ft/s. A valve at the end of the pipe is partially closed and the velocity instantly reduces to 4 ft/s. Estimate the surge pressure rise at the valve. Use a wave speed of 3000 ft/s for the water in the pipe.

Solution

From Equation (11.1):

$$\text{Pressure rise} = aV/g$$
$$= 3000 \times (6 - 4)/32.2 = 186.33 \text{ ft}$$

Example Problem 11.2

A steel pipe carries water at 20°C. The pipe diameter is 600 mm and wall thickness 6 mm. Assume Young's modulus E = 207 GPa and Poisson's ratio = 0.3. Use 1000 kg/m^3 for the density of water and a bulk modulus K = 2.2 GPa. Calculate the wave speed through this pipeline considering the three pipe constraints.

Solution

The D/t ratio is

$$600/6 = 100$$

therefore thin-walled pipe values for restraint factor C apply. Considering the pipe to be completely rigid, the wave speed is

$$a = (K/\rho)^{1/2} = (2.2 \times 10^9/1000)^{1/2} = 1483 \text{ m/s}$$

(a) For a pipe anchored at the upstream end only, from Equation (11.5):

$$C = 1 - 0.5\,\mu = 1 - 0.5 \times 0.3$$

or

$$C = 0.85$$

Therefore,

$$a = \frac{(K/\rho)^{1/2}}{(1 + CKD/Et)^{1/2}}$$
$$= \frac{1483}{[1 + 0.85 \times 2.2 \times 600/(207 \times 6)]^{1/2}}$$
$$= \frac{1483}{1.9034} = 1075 \text{ m/s}$$

(b) For a pipe anchored against any axial movement, from Equation (11.6):

$$C = 1 - \mu^2 = 1 - (0.3)^2 = 0.91$$

$$a = \frac{1483}{[1 + 0.91 \times 2.2 \times 600/(207 \times 6)]^{1/2}}$$

$$= 1057 \text{ m/s}$$

(c) For a pipe with each section anchored with expansion joints, from Equation (11.7):

$$C = 1$$

and

$$a = \frac{1483}{[1 + 1.0 \times 2.2 \times 600/(207 \times 6)]^{1/2}}$$
$$= 1033 \text{ m/s}$$

11.4 Transients in Cross-Country Pipelines

Cross-country pipelines and other trunk pipelines transporting crude oil, refined products, etc., are several hundred miles long and generally have multiple pump stations located along the pipeline. Average pump station spacing may be 30 to 60 miles or more. The pump stations are designed to provide the necessary pressure to overcome friction in the pipeline between pump stations and any static head lift necessary due to pipeline elevation differences.

The study of transient pressures in long-distance pipelines, called surge analysis, is fairly complex due to the fact that the frictional pressure drops in the pipelines are large compared with the instantaneous pressure surge due to sudden stoppage of flow.

A rigorous mathematical analysis of the pressure versus flow variations due to transient conditions results in partial differential equations that need to be solved with specified boundary conditions. Such solutions are difficult, if not impossible, using manual methods, even with a scientific calculator. However, over the years engineers and scientists have simplified these equations using finite difference methods and other mathematical approximations to result in the solution of simple simultaneous equations at every nodal point along the pipeline. Of course, the more the pipeline is subdivided, the higher is the accuracy of calculation. Today, with high-powered desktop computers, we can model these transient pressures in pipelines using the Method of Characteristics, which has proven to be quite accurate when compared with field tests. Several commercial computer programs are available to model transient pressures in pipelines due to valve closure, pump start-up and shut-down, and other upset conditions.

The potential surge is the instantaneous pressure rise due to sudden stoppage of a flow, such as closing of a valve at the end of the pipeline. This pressure was shown from Equation (11.1) to be

$$\Delta H = aV/g$$

The increase in storage capacity of a pipeline due to increase in pressure is termed line packing. Line packing occurs when the flow in a pipeline is changed by gradual or sudden closure of a valve. Consider a long crude oil pipeline flowing at a steady flow rate of Q bbl/hr. Due to sudden closure of a valve downstream there is a pressure rise immediately upstream of the valve. A pressure wave with an amplitude of ΔH travels from the valve upstream towards the pump station. Oil flows downstream of the wave front, pressure rises gradually at the downstream end, and the hydraulic grade line gradually reaches the horizontal level. During this time more oil gets packed between the valve and the pressure wave, hence the term "line pack." The pressure rise at the valve is made up of two components:

1. The potential surge pressure, ΔH, from Equation (11.1) due to instantaneous closure of the valve
2. The pressure rise due to line pack, ΔP

The pressure rise ΔP due to line pack in a long oil pipeline may be several times the size of the potential surge ΔH, depending on the length of the pipeline, its diameter, etc.

Attenuation is the process by which the amplitude of the surge pressure wave is reduced as it propagates along the pipeline due to the reduction in the velocity differential across the wave front. This is because, even though the wave front has passed a certain location on the pipeline, the oil continues to flow toward the valve. This causes a reduction in the velocity differential across the wave front as the wave propagates upstream.

In addition, the amplitude of the surge is also decreased due to pipe friction. This reduction, however, is a smaller component than that caused by the velocity differential across the wave front.

For a detailed analysis of transient analysis in long pipelines, see Wylie and Streeter in the References section.

11.5 Summary

In this chapter we have introduced the concept of unsteady flow in pipelines. If the velocity, flow rate, and pressure at any point in a pipeline change with time, unsteady flow is said to occur. Closing and opening valves, starting and stopping pumps, and injecting volumes or changing products all contribute to some form of unsteady flow. Some unsteady flow situations can cause pressure surges or transients that, when superimposed on the existing steady-state pressure gradient, may cause overpressure of a pipeline. Design codes for petroleum pipelines dictate that such transient overpressures cannot exceed 110% of the maximum allowable operating pressure in a pipeline.

We illustrated the concept of unsteady flow using a simple example of a tank connected to a pipe with a valve at the end. Sudden closure of the valve causes a pressure rise at the valve and generates a pressure wave that travels in the opposite direction of the flow at the speed of sound in the liquid. It was shown that flow rate and pressure variations occur throughout the pipeline due to the pressure wave traveling back and forth until it is attenuated by friction and equilibrium is established after a sufficient period of time. The wave speed depends mainly on the liquid bulk modulus and density. It also depends upon the pipe size and the elastic modulus of the pipe material.

Transient pressures developed in long-distance pipelines are generally modeled using computer programs based on the method of characteristics. This approach reduces the complex partial differential equations to simpler simultaneous equations which are solved sequentially from node to node, by subdividing the pipeline into equal segments. For more detailed analysis of unsteady flow the reader is referred to one of the publications listed in the References section.

11.6 Problems

11.6.1 Crude oil flows through a 16 in. diameter, 0.250 in. wall thickness steel pipeline, at a flow rate of 150,000 bbl/day. A valve located at the end of the pipeline is suddenly closed. Determine the potential surge generated and the wave speed in the pipeline assuming the pipe is anchored throughout against axial movement. Bulk modulus of crude oil = 280,000 psi, specific gravity = 0.89. Modulus of elasticity of steel = 30×10^6 psi. Poisson's ratio = 0.3.

11.6.2 Compare the wave speed in a rubber pipeline 250 mm in diameter, 6 mm wall thickness carrying water, with a similar pipeline 60 mm thick. $E = 0.1\,\text{GPa}$, $\mu = 0.45$, $K = 2.2\,\text{GPa}$, $\rho = 1000\,\text{kg/m}^3$. Assume the pipe is anchored at one end only.

11.6.3 A steel pipeline 12.75 in. in diameter and 0.250 in. wall thickness is used to transport water between two storage tanks. Calculate the wave speeds and restraint factor C using thin-walled and thick-walled formulas.

12

Pipeline Economics

In this chapter we discuss the cost of a pipeline and the various components that contribute to the economics of pipelines. These include the major components of the initial capital costs and the recurring annual costs. We also examine how the transportation charge is established based on throughput rates, project life, interest rate, and financing scenarios.

12.1 Economic Analysis

In any pipeline investment project we must perform an economic analysis of the pipeline system to ensure that we have the right equipment and materials at the right cost to perform the necessary service and provide a profitable income for the venture. The previous chapters helped determine the pipe size, pipe material, pumping equipment, etc., necessary to transport a given volume of a product. In this chapter we analyze the cost implications and how to decide on the economic pipe size and pumping equipment required to provide the optimum rate of return on investment.

The major capital components of a pipeline system consist of the pipe, pump stations, storage tanks, valves, fittings, and meter stations. Once this capital is expended, and the pipeline has been installed and the pump station and other facilities built, annual operating and maintenance costs for these facilities will be incurred. Annual costs will also include general and administrative (G&A) costs including payroll costs, rental and lease costs, and other recurring costs necessary for the safe and efficient operation of the pipeline system. The revenue for this operation will be in the form of pipeline tariffs collected from companies that ship products through this pipeline. The capital necessary for building this pipeline system may be partly owner equity and partly borrowed money. There will be investment hurdles and rate of return (ROR) requirements imposed by equity owners and financial institutions that lend the capital for the project. Regulatory requirements will also dictate the maximum revenue that may be collected and the ROR that may be realized, as transportation services. An economic analysis must be performed for the project taking into account all these factors and a reasonable project life of 20 to 25 years, or more in some cases.

These concepts will be illustrated by examples in subsequent sections of this chapter. Before we discuss the details of each cost component, it will be instructive and beneficial to calculate the transportation tariff and cost of service using a simple example.

Example Problem 12.1

Consider a new pipeline that is being built for transporting crude oil from a tank farm to a refinery. For the first phase (10 years) it is estimated that shipping volumes will be 100,000 bbl/day. Calculations indicate that a 16 in. pipeline 100 miles long with two pump stations will be required. The capital cost for all facilities is estimated to be $72 million. The annual operating cost including electrical power, O&M, G&A, etc., is estimated to be $5 million. The project is financed at a debt/equity ratio of 80/20. The interest rate on debt is 8% and the rate of return allowed by regulators is 12%. Consider a project life of 20 years and overall tax rate of 40%.

(a) What is the annual cost of service for this pipeline?
(b) Based on the fixed throughput rate of 100,000 bbl/day, and a load factor of 95%, what tariff rate can be charged within the regulatory guidelines?
(c) During the second phase, volumes are expected to increase by 20% (years 11 through 20). Estimate the revised tariff rate for the second phase assuming no capital cost changes to pump stations and other facilities. Use an increased annual operating cost of $7 million and the same load factor as before.

Solution

(a) The total capital cost of all facilities is \$72 million. We start by calculating the debt capital and equity capital based on the given 80/20 debt/equity ratio.

$$\text{Debt capital} = 0.80 \times 72 = \$57.6 \text{ million}$$

$$\text{Equity capital} = 72 - 57.6 = \$14.4 \text{ million}$$

The debt capital of \$57.6 million is borrowed from a bank or financial institution at the 8% annual interest rate. To retire this debt over the project life of 20 years, we must account for this interest payment in our annual costs.

In a similar manner, the equity investment of 14.4 million may earn 12% ROR. Since the tax rate is 40%, the annual cost component of the equity will be adjusted to compensate for the tax rate.

$$\text{Interest cost per year} = 57.6 \times 0.08 = \$4.61 \text{ million}$$

$$\text{Equity cost per year} = 14.4 \times 0.12/(1 - 0.4) = \$2.88 \text{ million}$$

Assuming a straight-line depreciation for 20 years, our yearly depreciation cost for \$72 million capital is calculated as follows:

$$\text{Depreciation per year} = 72/20 = \$3.6 \text{ million}$$

By adding annual interest expense, annual equity cost and annual depreciation cost to the annual O&M cost, we get the total annual service cost to operate the pipeline:

$$\text{Total cost of service} = 4.61 + 2.88 + 3.60 + 5.00 = \$16.09 \text{ million/yr}$$

(b) Based on the total cost of service of the pipeline, we can now calculate the tariff rate to be charged, spread over a flow rate of 100,000 bbl/day and a load factor of 95%:

$$\text{Tariff} = (16.09 \times 10^6)/(365 \times 100{,}000 \times 0.95) \text{ or } \$0.4640/\text{bbl}$$

(c) During phase 2, the new flow rate is 120,000 bbl/day and the total cost of service becomes:

$$\text{Total cost of service} = 4.61 + 2.88 + 3.60 + 7.00 = \$18.09 \text{ million/yr}$$

And the revised tariff becomes:

$$\text{Tariff} = (18.09 \times 10^6)/(365 \times 120{,}000 \times 0.95) \text{ or } \$0.4348/\text{bbl}$$

12.2 Capital Costs

The capital cost of a pipeline project consists of the following major components:

Pipeline
Pump stations
Tanks and manifold piping
Valves, fittings, etc.
Meter stations
SCADA and telecommunication
Engineering and construction management
Environmental and permitting
Right-of-way acquisition cost
Other project costs such as allowance for funds used during construction (AFUDC) and contingency

12.2.1 Pipeline Costs

The capital cost of a pipeline consists of material and labor for installation. To estimate the material cost we will use the following method:

$$\text{Pipe material cost} = 10.68(D - t)\,t \times 2.64 \times L \times \text{cost per ton}$$

or

$$PMC = 28.1952\,L\,(D - t)\,t\,(Cpt) \qquad (12.1)$$

where
PMC = Pipe material cost, $
L = Pipe length, miles
D = Pipe outside diameter, in.
t = Pipe wall thickness, in.
Cpt = Pipe cost, $/ton

In SI units, Equation (12.1) can be written as

$$PMC = 0.02463\,L\,(D - t)\,t\,(Cpt) \qquad (12.2)$$

where
PMC = Pipe material cost, $
L = Pipe length, km
D = Pipe outside diameter, mm
t = Pipe wall thickness, mm
Cpt = Pipe cost, $/metric ton

Since the pipe will be coated, wrapped, and delivered to the site, we will have to increase the material cost by some factor to account for these items or add the actual cost of these items to the pipe material cost.

Pipe installation cost or labor cost is generally stated in \$/ft or \$/mile of pipe. It may also be stated based on an inch-diameter-mile of pipe. Construction contractors will estimate the labor cost of installing a given pipeline based on a detailed analysis of the terrain, construction conditions, difficulty of access, and other factors. Historical data is available for estimating labor costs of various-size pipelines. In this section we will use approximate methods; these should be verified with contractors taking into account current labor rates, geographic and terrain issues. A good approach is to express the labor cost in terms of \$/inch diameter per mile of pipe. Thus we can say that a particular 16 in. pipeline can be installed at a cost of \$15,000 per inch-diameter-mile. Therefore, for a 100 mile, 16 in. pipeline we can estimate the labor cost to be

$$\text{Pipe labor cost} = \$15{,}000 \times 16 \times 100 = \$24 \text{ million}$$

Based on \$/ft cost this works out to be

$$\frac{24 \times 10^6}{100 \times 5280} = \$45.50 \text{ per ft}$$

Table A.14 in Appendix A shows a summary of typical labor costs. It must be noted that these numbers are approximate and must be verified with the contractor for a specific geographic location and level of construction difficulty.

In addition to labor costs for installing straight pipe, there may be other construction costs such as road crossings, railroad crossings, river crossings, etc. These are generally estimated as a lump sum for each item and added to the total pipe installation costs. For example there may be ten highway and road crossings totaling \$2 million and one major river crossing that may cost \$500,000. For simplicity, however, we will ignore these items for the present.

12.2.2 Pump Station

To estimate the pump station cost a detailed analysis would consist of preparing a material take-off from the pump station drawings and getting vendor quotes on major equipment such as pumps, drivers, switchgear, valves, instrumentation, etc., and estimating the station labor costs.

An approximate cost for pump stations can be estimated using a value for cost in dollars per installed horsepower. This is an all-inclusive number considering all facilities associated with the pump station. For example, we can use an installed cost of \$1500 per HP and estimate that a pump station with 5000 HP will cost

$$\$1500 \times 5000 = \$7.5 \text{ million}$$

In the above we used an all-inclusive number of \$1500 per installed HP. This figure takes into account all material and equipment cost and construction labor. Such values of installed cost per HP can be obtained from historical data on recently constructed pump stations. Larger HP pump stations will have smaller \$/HP costs while smaller pump stations with less HP will have a higher \$/HP cost, reflecting economies of scale.

12.2.3 Tanks and Manifold Piping

Tanks and manifold piping can be estimated fairly accurately by detailed material take-offs from construction drawings and from vendor quotes.

Generally, tank vendors quote installed tank costs in \$/bbl. Thus if we have a 50,000 bbl tank, it can be estimated at

$$50{,}000 \times \$10/\text{bbl} = \$500{,}000$$

based on an installed cost of \$10/bbl.

We would of course increase the total tankage cost by a factor of 10–20% to account for other ancillary piping and equipment.

As with installed HP costs, the unit cost for tanks decreases with tank size. For example, a 300,000 bbl tank may be based on \$6/bbl or \$8/bbl compared with the \$10/bbl cost for the smaller, 50,000 bbl tank.

12.2.4 Valves and Fittings

Valves and fittings may also be estimated as a percentage of the total pipe cost. However, if there are several mainline block valve locations that can be estimated as a lump sum cost, we can estimate the total cost of valves and fittings as follows: A typical 16 in. mainline block valve installation may cost \$100,000 per site including material and labor costs. If there are 10 such installations spaced 10 miles apart on a pipeline, we would estimate cost of valves and fittings to be \$1.0 million.

12.2.5 Meter Stations

Meter stations may be estimated as a lump sum fixed price for a complete site. For example, a 10 in. meter station with meter, valves, and piping instrumentation may be priced at \$250,000 per site including material and labor cost. If there are two such meter stations on the pipeline, we would estimate total meter costs at $500,000.

12.2.6 SCADA and Telecommunication System

This category covers costs associated with Supervisory Control and Data Acquisition (SCADA), telephone, microwave, etc. SCADA system costs include the facilities for remote monitoring, operation, and control of the pipeline from a central control center. Depending upon the length of the pipeline, number of pump stations, valve stations, etc., the cost of these facilities may range from $2 million to $5 million or more. An estimate based on the total project cost may range from 2% to 5%.

12.2.7 Engineering and Construction Management

Engineering and construction management consists of preliminary and detailed engineering design costs and personnel costs associated with management and inspection of the construction effort for pipelines, pump stations, and other facilities. This category usually ranges from 15% to 20% of total pipeline project costs.

12.2.8 Environmental and Permitting

In the past, environmental and permitting costs used to be a small percentage of the total pipeline system costs. In recent times, due to stricter environmental and regulatory requirements, this category now includes items such as an environmental impacts report, environmental studies pertaining to the flora and fauna, fish and game, endangered species, sensitive areas such as Native American burial sites, and allowance for habitat mitigation. The latter cost includes the acquisition of new acreage to compensate for areas disturbed by the pipeline route. This new acreage will then be allocated for parks, wildlife preserves, etc.

Permitting costs would include pipeline construction permits such as road crossings, railroad crossings, river and stream crossings, and permitting for antipollution devices for pump stations and tank farms.

Environmental and permitting costs may be as high as 10% to 15% of the total project costs.

12.2.9 Right-of-Way Acquisitions

Right of way (ROW) must be acquired for building a pipeline along private lands, farms, public roads, and railroads. In addition to initial acquisition costs there may be annual lease costs that the pipeline company will have to pay railroads, agencies, and private parties for pipeline easement and maintenance. The annual ROW costs would be considered an expense and would be included in the operating costs of the pipeline. For example, the ROW acquisitions costs for a pipeline project may be $20 million, which would be included in the total capital costs of the pipeline project. In addition, annual lease payments for ROW acquired may be a total of $500,000 a year, which would be included with other operating costs such as pipeline O&M, G&A costs, etc.

Historically, ROW costs have been in the range of 6% to 8% of total project costs for pipelines.

12.2.10 Other Project Costs

Other project costs would include allowance for funds used during construction (AFUDC), legal and regulatory costs, and contingency costs. Contingency costs cover unforeseen circumstances and design changes including pipeline rerouting to bypass sensitive areas, pump stations and facilities modifications not originally anticipated at the start of the project. AFUDC and contingency costs will range between 15% and 20% of the total project cost.

12.3 Operating Costs

The annual operating cost of a pipeline consists mainly of the following:

Pump station energy cost (electricity or natural gas)
Pump station equipment maintenance costs (equipment overhaul, repairs, etc.)
Pipeline maintenance cost including line rider, aerial patrol, pipe replacements, relocations, etc.
SCADA and telecommunication costs
Valve and meter station maintenance
Tank farm operation and maintenance
Utility costs: water, natural gas, etc.
Ongoing environmental and permitting costs
Right-of-way lease costs
Rentals and lease costs
General and administrative costs including payroll

In the above list, pump station costs include electrical energy and equipment maintenance costs, which can be substantial. Consider two pump stations of 5000 HP, each operating 24 hr a day, 350 days a year with a 2 week shutdown for maintenance. This can result in annual O&M costs of $6 million to $7 million based on electricity costs of 8 to 10 cents/kWh. In addition to the power cost other components of O&M costs include annual maintenance and overhead, which can range from \$0.50 million to \$1.0 million depending on the equipment involved.

12.4 Feasibility Studies and Economic Pipe Size

In many instances we have to investigate the technical and economic feasibility of building a new pipeline system to provide transportation services for liquids from a storage facility to a refinery or from a refinery to a tank farm. Other types of studies may include technical and economic feasibility studies for expanding the capacity of an existing pipeline system to handle additional throughput volumes due to increased market demand or refinery expansion.

Grass roots pipeline projects, where a brand new pipeline system needs to be designed from scratch, involve analysis of the best pipeline route, optimum pipe size, and pumping equipment required to transport a given volume of liquid. In this section we will learn how an economic pipe size is determined for a pipeline system, based on an analysis of capital and operating costs.

Consider a project in which a 100 mile pipeline is to be built to transport 8000 bbl/hr of refined products from a refinery to a storage facility. The question is: What pipe diameter and pump stations are optimal for handling this volume?

Let us assume that we selected a 16 in. diameter pipe to handle the designated volume and calculated that this system needs two pump stations of 2500 HP each. The total cost of this system of pipe and pump stations can be calculated and we will call this the 16 in. cost option. If we chose a 20 in. diameter pipeline, the design would require one 2000 HP pump station. In the first case, more HP and less pipe will be required. The 16 in. pipeline system would require approximately 20% less pipe than the 20 in. option. However, the 16 in. option requires 2.5 times the HP required in the 20 in. case. Therefore, the annual pump station operating costs for the 16 in. system would be higher than the 20 in. case, since the electric utility cost for the 5000 HP pump stations will be higher than that for the 2000 HP station required for the 20 in. system. Therefore, to determine the optimum pipe size required, we must analyze the capital costs and the annual operating costs to determine the scenario that gives us the least total cost, taking into account a reasonable project life. We would perform these calculations considering the time value of money, and select the option that results in the lowest present value (PV) of investment.

Generally, in any situation we must evaluate at least three or four different pipe diameters and calculate the total capital costs and operating costs for each pipe size. As indicated in the previous paragraph, the optimum pipe size and pump station configuration will be the alternative that minimizes total investment, after taking into account the time value of money over the life of the project. We will illustrate this using an example.

Example Problem 12.2

A city is proposing to build a 24 mile long pipeline to transport water at a flow rate of 14.4 million gal/day. There is static elevation head of 250 ft from the originating pump station to the delivery terminus. A minimum delivery pressure of 50 psi is required at the pipeline terminus.

The pipeline operating pressure must be limited to 1000 psi using steel pipe with a yield strength of 52,000 psi. Determine the optimum pipe diameter and the HP required for pumping this volume on a continuous basis, assuming 350 days operation, 24 hr a day. Electricity costs for driving the pumps will be based on 8 cents/kWh. The interest rate on borrowed money is 8% per year. Use the Hazen-Williams equation for pressure drop with a C-factor of 100.

Assume \$700/ton for pipe material cost and \$20,000/inch-diameter-mile for pipeline construction cost. For pump stations, assume a total installed cost of \$1500 per HP. To account for items other than pipe and pump stations in the total cost use a 25% factor.

Solution

First we have to bracket the pipe diameter range. If we consider 20 in. diameter pipe, 0.250 in. wall thickness, the average water velocity using Equation (3.11) will be

$$V = 0.4085 \times 10{,}000/19.5^2 = 10.7 \text{ ft/s}$$

where 10,000 gal/min is the flow rate based on 14.4 million gal/day. This is not a very high velocity. Therefore, 20 in. pipe can be considered as one of the options. We will compare this with two other pipe sizes: 22 in. and 24 in. nominal diameter.

Initially, we will assume 0.500 in. pipe wall thickness for the 22 in. and 24 in. pipes. Later we will calculate the actual required wall thickness for the given MAOP. Using ratios, the velocity in the 22 in. pipe will be approximately

$$10.7 \times (19.5/21)^2 \text{ or } 9.2 \text{ ft/s}$$

and the velocity in the 24 in. pipe will be

$$10.7 \times (19.5/23)^2 \text{ or } 7.7 \text{ ft/s}$$

Thus, the selected pipe sizes of 20 in., 22 in., and 24 in. will result in a water velocity between 7.7 ft/s and 10.7 ft/s, which is an acceptable range of velocities in a pipe.

Next we need to choose a suitable wall thickness for each pipe size to limit the operating pressure to 1000 psi. Using the internal design pressure Equation (4.3), we calculate the pipe wall thickness required as follows:

For 20 in. pipe, the wall thickness is

$$T = 1000 \times 20/(2 \times 52{,}000 \times 0.72) = 0.267 \text{ in.}$$

Similarly for the other two pipe sizes we calculate:
For 22 inch pipe, the wall thickness is

$$T = 1000 \times 22/(2 \times 52{,}000 \times 0.72) = 0.294 \text{ in.}$$

For 24 in. pipe, the wall thickness is

$$T = 1000 \times 24/(2 \times 52{,}000 \times 0.72) = 0.321 \text{ in.}$$

Using the closest commercially available pipe wall thicknesses, we choose the following three sizes:

20 in., 0.281 in. wall thickness (MAOP = 1052 psi)
22 in., 0.312 in. wall thickness (MAOP = 1061 psi)
24 in., 0.344 in. wall thickness (MAOP = 1072 psi)

The revised MAOP values for each pipe size, with the slightly higher than required minimum wall thickness, were calculated as shown within parentheses above. With the revised pipe wall thickness, the velocity calculated earlier will be corrected to

$$V_{20} = 10.81 \text{ ft/s}$$

$$V_{22} = 8.94 \text{ ft/s}$$

$$V_{24} = 7.52 \text{ ft/s}$$

Next, we calculate the pressure drop due to friction in each pipe size at the given flow rate of 10,000 gal/min, using the Hazen-Williams equation with a C-factor of 100. From Equation (3.36) for the 20 in. pipeline

$$10{,}000 \times 60 \times 24/42 = 0.1482 \times 100(20 - 2 \times 0.281)^{2.63}(Pm/1.0)^{0.54}$$

Rearranging and solving for the pressure drop P_m we get

$$P_m = 63.94 \text{ psi/mile for the 20 in. pipe}$$

Similarly, we get the following for the pressure drop in the 22 in. and 24 in. pipelines:

P_m= 40.25 psi/mile for the 22 in. pipe

P_m= 26.39 psi/mile for the 24 in. pipe

We can now calculate the total pressure required for each pipe size, taking into account the friction drop in the 24 mile pipeline and the elevation head of 250 ft along with a minimum delivery pressure of 50 psi at the pipeline terminus.

Total pressure required at the origin pump station is:

$(63.94 \times 24) + 250 \times 1.0/2.31 + 50 = 1692.79$ psi for the 20 in. pipe
$(40.25 \times 24) + 250 \times 1.0/2.31 + 50 = 1124.23$ psi for the 22 in. pipe
$(26.39 \times 24) + 250 \times 1.0/2.31 + 50 = 791.59$ psi for the 24 in. pipe

Since the MAOP of the pipeline is limited to 1000 psi, it is clear that we would need two pump stations for the 20 in. and 22 in. pipeline cases while one pump station would suffice for the 24 in. pipeline case.

The total BHP required for each case will be calculated from the above total pressure and the flow rate of 10,000 gal/min using Equation (5.18) assuming a pump efficiency of 80%. We will also assume that the pumps require a minimum suction pressure of 50 psi.

$\text{BHP} = 10{,}000 \times (1693 - 50)/(0.8 \times 1714) = 11{,}983$ for 20 in.
$\text{BHP} = 10{,}000 \times (1124 - 50)/(0.8 \times 1714) = 7833$ for 22 in.
$\text{BHP} = 10{,}000 \times (792 - 50)/(0.8 \times 1714) = 5412$ for 24 in.

Increasing the BHP values above by 10% for installed HP and choosing the nearest motor size, we will use 14,000 HP for the 20 in. pipeline system, 9000 HP for the 22 in. system, and 6000 HP for the 24 in. pipeline system. If we had factored in a 95% efficiency for the electric motor and picked the next nearest size motor we would have arrived at the same HP motors as above.

To calculate the capital cost of facilities, we will use $700 per ton for steel pipe, delivered to the construction site. The labor cost for installing the pipe will be based on $20,000 per inch-diameter-mile.

The installed cost for pump stations will be assumed to be $1500 /HP. To account for other cost items discussed earlier in this chapter, we will add 25% to the subtotal of pipeline and pump station cost.

The estimated capital costs for the three pipe sizes are summarized in Table 12.1. Based on total capital costs alone, it can be seen that the 24 in. system is the best. However, we will have to look at the operating costs as well, before making a decision on the optimum pipe size.

Next, we calculate the operating cost for each scenario, using electrical energy costs for pumping. As discussed in an earlier section of this chapter,

Table 12.1 Capital Costs for Varying Pipe Sizes

	Pipe size		
Capital cost (million \$)	20 in.	22 in.	24 in.
Pipeline	2.62	3.21	3.85
Pump stations	21.00	13.50	9.00
Other (25%)	5.91	4.18	3.21
Total	29.53	20.88	16.07

many other items enter into the calculation of annual operating costs, such as O&M, G&A costs, etc. For simplicity, we will increase the electrical cost of the pump stations by a factor to account for all other operating costs.

Using the BHP calculated at each pump station for the three cases and 8 cents/kWh for electricity cost, we find that the annual operating cost for 24 hr operation per day, 350 days per year the annual costs are:

$11{,}983 \times 0.746 \times 24 \times 350 \times 0.08 = \$6.0$ million/yr for 20 in.
$7833 \times 0.746 \times 24 \times 350 \times 0.08 = \$3.93$ million/yr for 22 in.
$5412 \times 0.746 \times 24 \times 350 \times 0.08 = \$2.71$ million/yr for 24 in.

Strictly speaking, the above costs will have to be increased to account for the demand charge for starting and stopping electric motors. The utility company may charge based on the kW rating of the motor. This will range from \$4 to \$6 per kW/month. Using an average demand charge of \$5/kW/month, we get the following demand charges for the pump station in a 12 month period:

$14{,}000 \times 5 \times 12 = \$840{,}000$ for 20 in.
$9000 \times 5 \times 12 = \$540{,}000$ for 22. in.
$6000 \times 5 \times 12 = \$360{,}000$ for 24 in.

Adding the demand charges to the previously calculated electric power cost, we get the total annual electricity costs as

\$6.84 million/yr for 20 in.
\$4.47 million/yr for 22 in.
\$3.07 million/yr for 24 in.

Increasing above numbers by a 50% factor to account for other operating costs such as O&M, G&A, etc., we get the following for total annual costs for each scenario:

\$10.26 million/yr for 20 in.
\$6.7 million/yr for 22 in.

$4.6 million/yr for 24 in.

Next, we use a project life of 20 years and interest rate of 8% to perform a discounted cash flow (DCF) analysis, to obtain the present value of these annual operating costs. Then the total capital cost calculated earlier and listed in Table 12.1 will be added to the present values of the annual operating costs. The present value (PV) will then be obtained for each of the three scenarios.

PV of 20 in. system = \$29.53 + present value of \$10.26 million/yr at 8% for 20 years

or

$$PV_{20} = 29.53 + 100.74 = \$130.27 \text{ million}$$

Similarly, for the 22 in. and 24 in. systems, we get

$$PV_{22} = 20.88 + 65.78 = \$86.66 \text{ million}$$
$$PV_{24} = 16.07 + 45.16 = \$61.23 \text{ million}$$

Thus, based on the net present value of investment, we can conclude that the 24 in. pipeline system with one 6000 HP pump station is the preferred choice.

In the preceding calculations we made several assumptions for the sake of simplicity. We considered major cost components, such as pipeline and pump station costs, and added a percentage of the subtotal to account for other costs. Also, in calculating the PV of the annual costs we used constant numbers for each year. A more rigorous approach would require the annual costs be inflated by some percentage every year to account for inflation and cost of living adjustments. The Consumer Price Index (CPI) could be used in this regard. As far as capital costs go, we can get more accurate results if we perform a more detailed analysis of the cost of valves, meters, and tanks instead of using a flat percentage of the pipeline and pump station costs. The objective in this chapter was to introduce the reader to the importance of economic analysis and to outline a simple approach to selecting the economical pipe size.

In addition, the earlier section on cost of services and tariff calculations provided an insight into how transportation companies finance a project and collect revenues for their services.

12.5 Summary

We have reviewed the major cost components of a pipeline system consisting of pipe, pump station, etc., and illustrated methods of estimating the capital costs of these items. The annual costs such as electrical energy,

O&M, etc., were also identified and calculated for a typical pipeline. Using the capital cost and operating cost, the annual cost of service was calculated based on specified project life, interest cost, etc. We thus determined the transportation tariff that could be charged for shipments through the pipelines. Also a methodology for determining the optimal pipe size for a particular application using present value (PV) was explained. Considering three different pipe sizes, we determined the best option based on a comparison of PV of the three different cases.

12.6 Problems

12.6.1 Calculate the annual cost of service and transportation tariff to be charged for shipments through a refined products pipeline as follows. The pipeline is 90 km long, 400 mm diameter and 8 mm wall thickness and constructed of steel with a yield strength of 448 MPa. The terrain is essentially flat. It is used to transport gasoline (specific gravity = 0.74 and viscosity = 0.65 cSt at 60°F) at a flow rate of 2000 m^3/hr. The annual operating costs such as O&M, G&A, etc., are estimated to be $2 million and do not include power costs. Assume a power cost of $0.10 per kWh for the electric motor-driven pumps. The project will be financed at a debt/equity ratio 70/30. The interest rate on debt is 7% and the rate of return allowed by regulators is 14%. Assume a project life of 25 years and overall tax rate of 35%.

12.6.2 Compare pipeline sizes of 12 in., 14 in., and 16 in. for an application that requires shipments of crude oil from a tank farm to a refinery 30 miles away at 5000 bbl/hr. The tank farm is at an elevation of 350 ft while the refinery is at 675 ft elevation. The pipe MAOP is limited to 1400 psi. Consider 5LX-65 pipe and a 72% design factor. The suction pressure at the tank farm may be assumed at 50 psi and the delivery pressure at the refinery is 30 psi. The pipeline will be operated 355 days a year, 24 hr per day. Electricity cost is 6 cents/kWh.

12.6.3 In Problem 12.6.2, assuming the steady-state flow rate of 5000 bbl/hr remains constant for the first 10 years, what is the estimated tariff? Determine the reduction in tariff if the flow rate is increased by 20% for the next 10 years.

Appendix A

Tables and Charts

Table A.1 Units and Conversions

Item	English units	SI units	Conversion: English to SI
Mass	Slugs (slugs)	Kilogram (kg)	1 lb = 0.45359 kg
	Pound mass (lbm)		1 slug = 14.594 kg
	U.S. tons	Metric tonne (t)	1 U.S. ton = 0.9072 t
	Long tons		1 long ton = 1.016 t
Length	Inches (in.)	Millimeter (mm)	1 in. = 25.4 mm
	Feet (ft)	Meter (m)	1 ft = 0.3048 m
	Miles (mi)	Kilometer (km)	1 mile = 1.609 km
Area	Square feet (ft^2)	Square meter (m^2)	1 ft^2 = 0.0929 m^2
Volume	Cubic inch ($in.^3$)	Cubic millimeter (mm^3)	1 $in.^3$ = 16387.0 mm^3
	Cubic feet (ft^3)	Cubic meter (m^3)	1 ft^3 = 0.02832 m^3
	U.S. gallons (gal)	Liter (L)	1 gal = 3.785 L
	Barrel (bbl)		1 bbl = 42 U.S. gal
Density	Slugs per cubic foot ($slug/ft^3$)	Kilogram per cubic meter (kg/m^3)	1 $slug/ft^3$ = 515.38 kg/m^3
Specific weight	Pound per cubic foot (lb/ft^3)	Newton per cubic meter (N/m^3)	1 lb/ft^3 = 157.09 N/m^3
Viscosity (kinematic)	Square foot per second (ft^2/s)	m^2/s	1 ft^2/s = 0.092903 m^2/s
Flow rate	Gallon/minute (gal/min)	Liter/minute (L/min)	1 gal/min = 3.7854 L/min

	Barrel/hour (bbl/hr)	Cubic meter/hour (m^3/hr)	1 bbl/hr = 0.159 m^3/hr
	Barrel/day (bbl/day)		
Force	Pound (lb)	Newton (N)	1 lb = 4.4482 N
Pressure	Pound/square inch (psi)	Kilopascal (kPa)	1 psi = 6.895 kPa
	Pound per square inch (lb/in.^{2})	Kilogram/square centimeter (kg/cm^2)	1 psi = 0.0703 kg/cm^2
Velocity	Foot/second (ft/s)	Meter/second (m/s)	1 ft/s = 0.3048 m/s
Work and energy	British Thermal Units (Btu)	Joule (J)	1 Btu = 1055.0 J
Power	Btu/hour	Watt (W)	1 Btu/hr = 0.2931 W
		Joule/second (J/s)	
	Horsepower (HP)	Kilowatt (kW)	1 HP = 0.746 kW
Temperature	Degree Fahrenheit (°F)	Degree Celsius (°C)	1°F = 9/5°C + 32
	Degree Rankin (°R)	Degree Kelvin (K)	1°R = °F + 460
			1 K = °C + 273
Thermal conductivity	Btu/hr/ft/°F	W/m/°C	1 Btu/hr/ft/°F = 1.7307 W/m/°C
Specific heat	Btu/lb/°F	kJ/kg/°C	1 Btu/lb/°F = 4.1869 kJ/kg/°C

Table A.2 Common Properties of Petroleum Fluids

Product	Viscosity (cSt at 60°F)	API gravity	Specific gravity at 60°F	Reid vapor pressure (psi)
Regular gasoline				
Summer grade	0.70	62.0	0.7313	9.5
Interseasonal grade	0.70	63.0	0.7275	11.5
Winter grade	0.70	65.0	0.7201	13.5
Premium gasoline				
Summer grade	0.70	57.0	0.7467	9.5
Interseasonal grade	0.70	58.0	0.7165	11.5
Winter grade	0.70	66.0	0.7711	13.5
No. 1 fuel oil	2.57	42.0	0.8155	
No. 2 fuel oil	3.90	37.0	0.8392	
Kerosene	2.17	50.0	0.7796	
Jet fuel JP-4	1.40	52.0	0.7711	2.7
Jet fuel JP-5	2.17	44.5	0.8040	

Table A.3 Specific Gravity and API Gravity

Liquid	Specific gravity at 60°F	API gravity at 60°F
Propane	0.5118	N/A
Butane	0.5908	N/A
Gasoline	0.7272	63.0
Kerosene	0.7796	50.0
Diesel	0.8398	37.0
Light crude	0.8348	38.0
Heavy crude	0.8927	27.0
Very heavy crude	0.9218	22.0
Water	1.0000	10.0

Table A.4 Viscosity Conversions

Viscosity (SSU)	Viscosity (cSt)	Viscosity (SSF)
31.0	1.00	
35.0	2.56	
40.0	4.30	
50.0	7.40	
60.0	10.30	
70.0	13.10	12.95
80.0	15.70	13.70
90.0	18.20	14.44
100.0	20.60	15.24
150.0	32.10	19.30
200.0	43.20	23.50
250.0	54.00	28.0
300.0	65.00	32.5
400.0	87.60	41.9
500.0	110.00	51.6
600.0	132.00	61.4
700.0	154.00	71.1
800.0	176.00	81.0
900.0	198.00	91.0
1000.0	220.00	100.7
1500.0	330.00	150
2000.0	440.00	200
2500.0	550.00	250
3000.0	660.00	300
4000.0	880.00	400
5000.0	1100.00	500
6000.0	1320.00	600
7000.0	1540.00	700
8000.0	1760.00	800
9000.0	1980.00	900
10000.0	2200.00	1000
15000.0	3300.00	1500
20000.0	4400.00	2000

Table A.5 Thermal Conductivities

Substance	Thermal conductivity (Btu/hr/ft/°F)
Fire clay brick (burnt at 2426°F)	0.60 to 0.63
Fire clay brick (burnt at 2642°F)	0.74 to 0.81
Fire clay brick (Missouri)	0.58 to 1.02
Portland cement	0.17
Mortar cement	0.67
Concrete	0.47 to 0.81
Cinder concrete	0.44
Glass	0.44
Granite	1.0 to 2.3
Limestone	0.73 to 0.77
Marble	1.6
Sandstone	0.94 to 1.2
Corkboard	0.025
Fiber insulating board	0.028
Aerogel, silica	0.013
Coal, anthracite	0.15
Coal, powdered	0.067
Ice	1.28
Sandy soil, dry	0.25 to 0.40
Sandy soil, moist	0.50 to 0.60
Sandy soil, soaked	1.10 to 1.30
Clay soil, dry	0.20 to 0.30
Clay soil, moist	0.40 to 0.50
Clay soil, moist to wet	0.60 to 0.90
River water	2.00 to 2.50
Air	2.00

Table A.6 Absolute Roughness of Pipe

Pipe material	Roughness (mm)	Roughness (in.)
Riveted steel	0.9 to 9.0	0.0354 to 0.354
Concrete	0.3 to 3.0	0.0118 to 0.118
Wood stave	0.18 to 0.9	0.0071 to 0.0354
Cast iron	0.26	0.0102
Galvanized iron	0.15	0.0059

(*Continued*)

Table A.6 Continued

Pipe material	Roughness (mm)	Roughness (in.)
Asphalted cast iron	0.12	0.0047
Commercial steel	0.045	0.0018
Wrought iron	0.045	0.0018
Drawn tubing	0.0015	0.000059

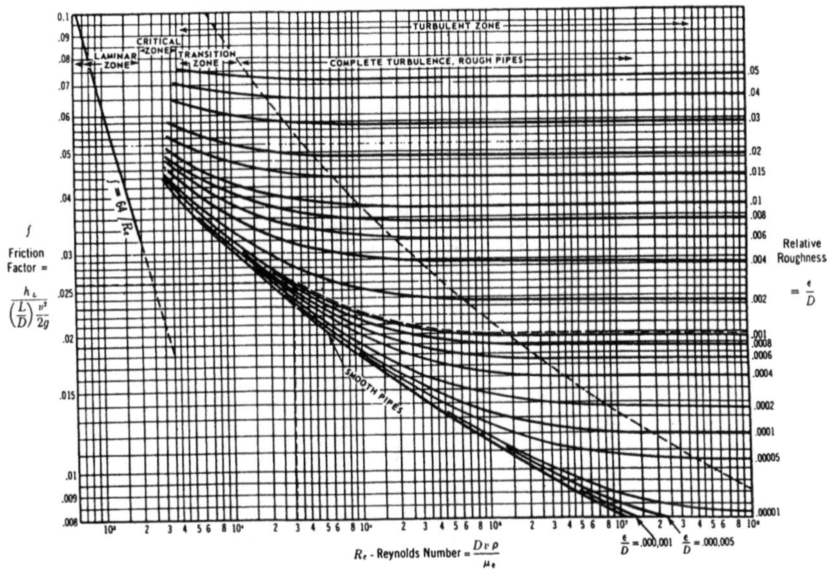

Figure A.7 Moody diagram.

Table A.8 Typical Hazen-Williams C-Factors

Pipe material	C-factor
Smooth pipes (all metals)	130–140
Smooth wood	120
Smooth masonry	120
Vitrified clay	110
Cast iron (old)	100
Iron (worn/pitted)	60–80
Polyvinyl chloride (PVC)	150
Brick	100

Table A.9 Friction Loss in Valves: Resistance Coefficient K

Description	L/D	Nominal pipe size (in.)											
		1/2	3/4	1.0	1¼	1½	2	2½ to 3	4	6	8 to 10	12 to 16	18 to 24
Gate valve	8	0.22	0.20	0.18	0.18	0.15	0.15	0.14	0.14	0.12	0.11	0.10	0.10
Globe valve	340	9.2	8.5	7.8	7.5	7.1	6.5	6.1	5.8	5.1	4.8	4.4	4.1
Ball valve	3	0.08	0.08	0.07	0.07	0.06	0.06	0.05	0.05	0.05	0.04	0.04	0.04
Butterfly valve							0.86	0.81	0.77	0.68	0.63	0.35	0.30
Plug valve, straightway	18	0.49	0.45	0.41	0.40	0.38	0.34	0.32	0.31	0.27	0.25	0.23	0.22
Plug valve, three-way thru-flo	30	0.81	0.75	0.69	0.66	0.63	0.57	0.54	0.51	0.45	0.42	0.39	0.36
Plug valve, branch-flo	90	2.43	2.25	2.07	1.98	1.89	1.71	1.62	1.53	1.35	1.26	1.17	1.08

Table A.10 Equivalent Lengths of Valves and Fittings

Description	L/D
Gate valve	8
Globe valve	340
Ball valve	3
Swing check valve	50
Standard elbow, 90°	30
Standard elbow, 45°	16
Long-radius elbow, 90°	16

Example: A 14-in. gate valve has L/D ratio = 8. Therefore equivalent length = 8 × 14 in. = 112 in. = 9.25 feet.

Table A.11 Seam Joint Factors for Pipes

Specification	Pipe class	Seam joint factor (E)
ASTM A53	Seamless	1.00
	Electric resistance welded	1.00
	Furnace lap welded	0.80
	Furnace butt welded	0.60
ASTM A106	Seamless	1.00
ASTM A134	Electric fusion arc welded	0.80
ASTM A135	Electric resistance welded	1.00
ASTM A139	Electric fusion welded	0.80
ASTM A211	Spiral welded	0.80
ASTM A333	Seamless	1.00
	Welded	1.00
ASTM A381	Double submerged arc welded	1.00
ASTM A671	Electric fusion welded	1.00
ASTM A672	Electric fusion welded	1.00
ASTM A691	Electric fusion welded	1.00
API 5L	Seamless	1.00
	Electric resistance welded	1.00
	Electric flash welded	1.00
	Submerged arc welded	1.00
	Furnace lap welded	0.80
	Furnace butt welded	0.60
API 5LX	Seamless	1.00
	Electric resistance welded	1.00
	Electric flash welded	1.00
	Submerged arc welded	1.00
API 5LS	Electric resistance welded	1.00
	Submerged arc welded	1.00

Table A.12 ANSI Pressure Ratings

Class	Allowable pressure (psi)
150	275
300	720
400	960
600	1440
900	2160
1500	3600

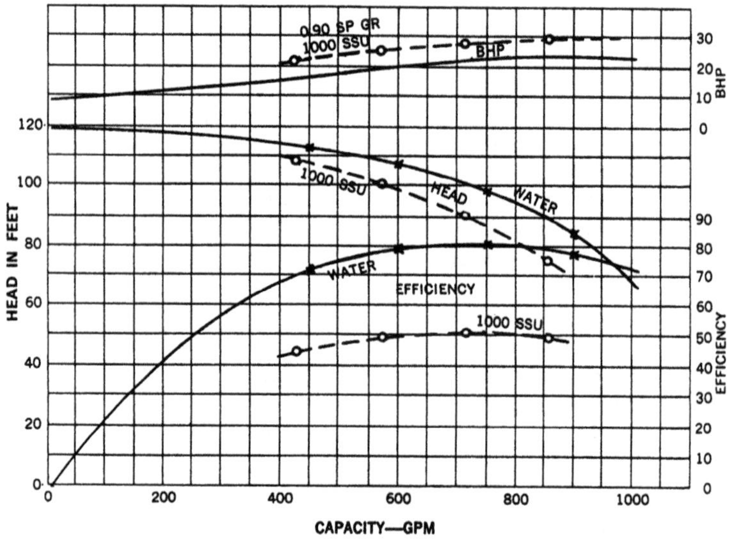

SAMPLE CALCULATIONS	$0.6 \times Q_{N\,w}$	$0.8 \times Q_{N\,w}$	$1.0 \times Q_{N\,w}$	$1.2 \times Q_{N\,w}$
Water capacity (Q_w)	450	600	750	900
Water head in feet (H_w)	114	108	100	86
Water efficiency (E_w)	72.5	80	82	79.5
Viscosity of liquid	1000 SSU	1000 SSU	1000 SSU	1000 SSU
C_Q—from chart	0.95	0.95	0.95	0.95
C_H—from chart	0.96	0.94	0.92	0.89
C_E—from chart	0.635	0.635	0.635	0.635
Viscous capacity—$Q_w \times C_Q$	427	570	712	855
Viscous head—$H_w \times C_H$	109.5	101.5	92	76.5
Viscous efficiency—$E_w \times C_E$	46.0	50.8	52.1	50.5
Specific gravity of liquid	0.90	0.90	0.90	0.90
bhp viscous	23.1	25.9	28.6	29.4

Figure A.13 Centrifugal pump performance viscosity correction. (Courtesy of Hydraulic Institute, Parsippany, NJ; www.pumps.org.)

Table A.14 Approximate Pipeline Construction Cost

Pipe diameter (in.)	Average cost ($/in.-dia/mile)
8	18,000
10	20,000
12	22,000
16	14,900
20	20,100
24	33,950
30	34,600
36	40,750

Table A.15 Thermal Hydraulics Summary

********** TEMPERATURE AND PRESSURE PROFILE **********

Distance (mi)	FlowRate (bbl/day)	Temperature (degF)	SpGrav --	Viscosity CST	Pressure (psi)	MAOP (psi)	Location
0.00	100,000.00	190.00	0.9355	68.72	50.00	1400.00	Compton
0.00	100,000.00	192.74	0.9345	63.93	1181.18	1400.00	Compton
18.00	100,000.00	186.76	0.9366	74.96	792.70	1400.00	
22.00	100,000.00	185.50	0.9371	77.59	298.41	1400.00	
23.00	100,000.00	185.19	0.9372	78.26	299.60	1400.00	
24.00	100,000.00	184.88	0.9373	78.93	300.74	1400.00	
29.00	100,000.00	183.35	0.9378	82.36	306.20	1400.00	
35.00	100,000.00	181.56	0.9385	86.60	248.57	1400.00	
40.00	100,000.00	180.12	0.9390	90.24	50.00	1400.00	Dimpton
40.00	100,000.00	183.34	0.9378	82.39	1371.78	1400.00	Dimpton
40.58	100,000.00	183.16	0.9379	82.79	1348.95	1400.00	
50.00	100,000.00	180.38	0.9389	89.56	977.68	1400.00	
54.00	100,000.00	179.25	0.9393	92.52	868.04	1400.00	
55.00	100,000.00	178.97	0.9394	93.28	865.82	1400.00	
60.00	100,000.00	177.58	0.9398	97.10	813.82	1400.00	
70.00	100,000.00	174.91	0.9408	105.06	666.49	1400.00	
80.00	100,000.00	172.37	0.9417	113.38	391.64	1400.00	
90.00	100,000.00	169.97	0.9425	122.06	213.27	1400.00	
100.00	100,000.00	167.69	0.9433	131.08	50.00	1400.00	Terminus

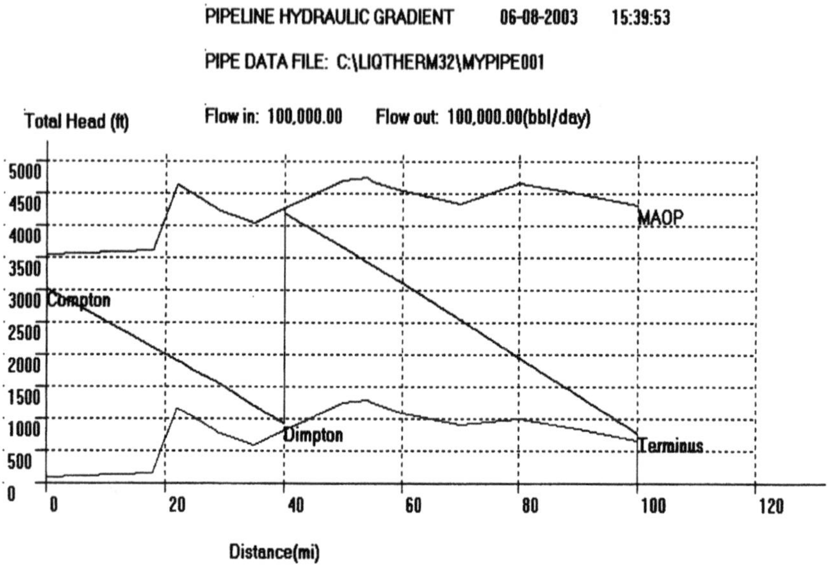

Figure A.16 Thermal hydraulics pressure gradient.

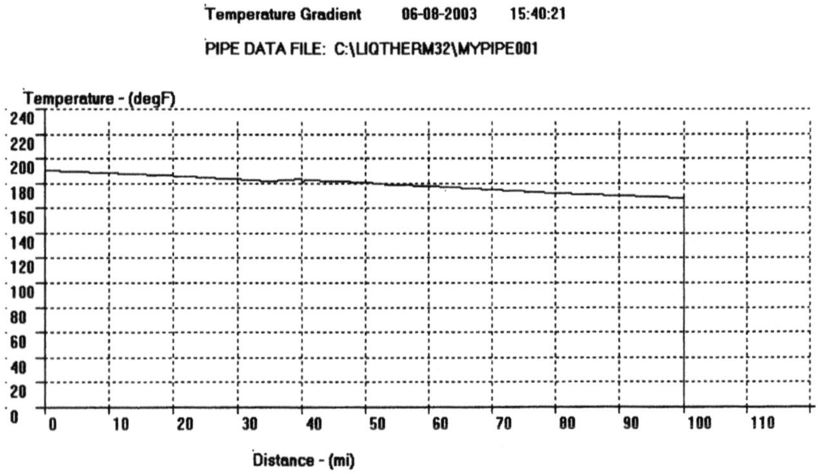

Figure A.17 Thermal hydraulics temperature gradient.

Appendix B

Answers to Selected Problems

Chapter 2

Problem 2.9.1: Specific weight = 52.88 lb/ft^3. Specific gravity = 0.8474.
Problem 2.9.2: Specific gravity at 85°F = 0.845.
Problem 2.9.3: Specific gravity = 0.7428.
Problem 2.9.7: Blended viscosity = 145 SSU.
Problem 2.9.8: 75% of liquid A and 25% of liquid B.

Chapter 3

Problem 3.15.1: Average velocity = 8.03 ft/s. Reynolds number = 203,038.
Problem 3.15.2: Darcy friction factor f = 0.0164. Modified Colebrook-White friction factor = 0.0167. Pressure drop = 101.66 psi.
Problem 3.15.4: Total pressure = 36.72 MPa. Four pump stations.

Chapter 4

Problem 4.5.2: MAOP = 1380 psi. Hydrotest range = 1725 to 1820 psi.
Problem 4.5.4: Volume per mile = 926.19 bbl.

Chapter 5

Problem 5.13.1: (a) 914 psi. (b) BHP = 2201; motor HP = 2293 HP. (c) 2500 HP. (d) 6510 bbl/hr.

Problem 5.13.2: 640 mm diameter, 10 mm wall thickness.

Problem 5.13.3: 8177 bbl/hr gasoline flow rate, gasoline head = 4277 ft; diesel head = 3724 ft.

Chapter 6

Problem 6.7.1: (a) 0.375 in. wall thickness. (b) Two pump stations. (c) Two additional pump stations. (d) 6888 HP and 18,530 HP at 75% pump efficiency.

Chapter 7

Problem 7.14.1: (a) Trim impeller in pump A or pump B. (b) 9.2 in. impeller for pump A. (c) 3282 RPM.

Problem 7.14.2:

Q (gal/min):	0	500	1000	1500	2000
P (psi):	130	208	419	758	1224

Chapter 8

Problem 8.5.1: (a) 122 psi. (b) $111,542 based on 350 days operation/yr.

Problem 8.5.2: Speed of VSD pump = 3126 RPM.

Chapter 9

Problem 9.4.1: Minimum flow rate = 1050 bbl/hr; HP = 298.

Problem 9.4.2: Minimum flow rate = 1500 bbl/hr; HP = 895.

Chapter 10

Problem 10.9.1: (a) R = 212,750. (b) 3.84 in. throat diameter. (c) R = 221,288.

Problem 10.9.3: Flow rate = 959 gal/min.

Chapter 11

Problem 11.6.1: Potential surge = 903 ft. Wave speed = 3913 ft/s.
Problem 11.6.3:

Thin wall:	3967 ft/s	3926 ft/s	3868 ft/s
Thick wall:	3942 ft/s	3903 ft/s	3847 ft/s

Appendix C

Summary of Formulas

Chapter 2

API Gravity/Specific Gravity

$$\text{Specific gravity (Sg)} = 141.5/(131.5 + \text{API}) \tag{2.1}$$

$$\text{API} = 141.5/\text{Sg} - 131.5 \tag{2.2}$$

Specific Gravity Versus Temperature

$$S_T = S_{60} - a(T - 60) \tag{2.3}$$

where

S_T = Specific gravity at temperature T
S_{60} = Specific gravity at 60°F
T = Temperature, °F
a = A constant that depends on the liquid

Specific Gravity Blending

$$S_b = \frac{(Q_1 \times S_1) + (Q_2 \times S_2) + (Q_3 \times S_3) + \cdots}{Q_1 + Q_2 + Q_3 + \cdots} \tag{2.4}$$

where

S_b = Specific gravity of the blended liquid
Q_1, Q_2, Q_3, etc. = Volume of each component
S_1, S_2, S_3, etc. = Specific gravity of each component

Viscosity Conversion

$$\text{Centistokes} = 0.226(\text{SSU}) - 195/(\text{SSU}) \quad \text{for } 32 \le \text{SSU} \le 100 \quad (2.8)$$
$$\text{Centistokes} = 0.220(\text{SSU}) - 135/(\text{SSU}) \quad \text{for SSU} > 100 \quad (2.9)$$
$$\text{Centistokes} = 2.24(\text{SSF}) - 184/(\text{SSF}) \quad \text{for } 25 < \text{SSF} = 40 \quad (2.10)$$
$$\text{Centistokes} = 2.16(\text{SSF}) - 60/(\text{SSF}) \quad \text{for SSF} > 40 \quad (2.11)$$

Viscosity Versus Temperature

$$\text{Log Log}(Z) = A - B\,\text{Log}(T) \quad (2.15)$$

where

Log = logarithm to base 10
Z depends on the viscosity of the liquid
ν = viscosity of liquid, cSt
T = Absolute temperature, °R or K

A and B are constants that depend on the specific liquid. The variable Z is defined as follows:

$$Z = (\nu + 0.7 + C - D) \quad (2.16)$$

$$C = \exp[-1.14883 - 2.65868(\nu)] \quad (2.17)$$

$$D = \exp[-0.0038138 - 12.5645(\nu)] \quad (2.18)$$

Viscosity Blending

$$\sqrt{V_b} = \frac{Q_1 + Q_2 + Q_3 + \cdots}{(Q_1/\sqrt{V_1}) + (Q_2/\sqrt{V_2}) + (Q_3/\sqrt{V_3})} \quad (2.21)$$

where

V_b = Viscosity of blend, SSU
Q_1, Q_2, Q_3, etc. = Volumes of each component
V_1, V_2, V_3, etc. = Viscosity of each component, SSU

$$H = 40.073 - 46.414\,\text{Log}_{10}\text{Log}_{10}(V + B) \quad (2.22)$$
$$B = 0.931(1.72)^V \quad \text{for } 0.2 < V < 1.5 \quad (2.23)$$
$$B = 0.6 \quad \text{for } V \ge 1.5 \quad (2.24)$$
$$Hm = [H1(\text{pct}1) + H2(\text{pct}2) + H3(\text{pct}3) + \cdots]/100 \quad (2.25)$$

where

H, H1, H2, etc. = Blending Index of liquids
Hm = Blending Index of mixture
B = Constant in Blending Index equation
V = Viscosity, cSt
pct1, pct2, etc. = Percentage of liquids 1, 2, etc., in blended mixture

Bulk Modulus

Adiabatic bulk modulus:

$$Ka = A + B(P) - C(T)^{1/2} - D(API) - E(API)^2 + F(T)(API) \qquad (2.27)$$

where

$A = 1.286 \times 10^6$
$B = 13.55$
$C = 4.122 \times 10^4$
$D = 4.53 \times 10^3$
$E = 10.59$
$F = 3.228$
P = Pressure, psig
T = Temperature, °R
API = API gravity of liquid

Isothermal Bulk Modulus:

$$Ki = A + B(P) - C(T)^{1/2} + D(T)^{3/2} - E(API)^{3/2} \qquad (2.28)$$

where

$A = 2.619 \times 10^6$
$B = 9.203$
$C = 1.417 \times 10^5$
$D = 73.05$
$E = 341.0$
P = Pressure, psig
T = Temperature, °R
API = API gravity of liquid

Bernoulli's Equation

$$Z_A + P_A/\gamma + V_A^2/2g + H_P = Z_B + P_B/\gamma + V_B^2/2g + \Sigma h_L \qquad (2.37)$$

where

Z_A, P_A and V_A = Elevation, pressure, and liquid velocity at point A
Z_B, P_B and V_B = Elevation, pressure, and liquid velocity at point B
γ = Specific weight of the liquid

H_P = Pump head input at point A
h_L = Head lost in friction between point A and point B

Chapter 3

Pressure and Head

$$\text{Head} = 2.31(\text{psig})/\text{Spgr} \quad \text{ft(English units)} \tag{3.7}$$

$$\text{Head} = 0.102(\text{kPa})/\text{Spgr} \quad \text{m(SI units)} \tag{3.8}$$

where
Spgr = Liquid specific gravity

Velocity of Flow

$$V = 0.0119(\text{bbl/day})/D^2 \tag{3.10}$$

$$V = 0.4085(\text{gal/min})/D^2 \tag{3.11}$$

$$V = 0.2859(\text{bbl/hr})/D^2 \tag{3.12}$$

where
V = Velocity, ft/s
D = Inside diameter, in.

$$V = 353.6777(\text{m}^3/\text{hr})/D^2 \tag{3.13}$$

where
V = Velocity, m/s
D = Inside diameter, mm

Reynolds Number

$$R = VD\rho/\mu \tag{3.14}$$

$$R = VD/\nu \tag{3.15}$$

where
V = Average velocity, ft/s
D = Pipe internal diameter, ft
ρ = Liquid density, slugs/ft^3
μ = Absolute viscosity, lb-s/ft^2
R = Reynolds number, dimensionless
ν = Kinematic viscosity, ft^2/s

$$R = 92.24\ Q/(\nu D) \tag{3.16}$$

where

Q = Flow rate, bbl/day
D = Internal diameter, in.
ν = Kinematic viscosity, cSt

$$R = 3160\ Q/(\nu D) \tag{3.17}$$

where

Q = Flow rate, gal/min
D = Internal diameter, in.
ν = Kinematic viscosity, cSt

$$R = 353{,}678\ Q/(\nu D) \tag{3.18}$$

where

Q = Flow rate, m^3/h
D = Internal diameter, mm
ν = Kinematic viscosity, cSt

Darcy-Weisbach Equation for Head Loss

$$h = f(L/D)(V^2/2g) \tag{3.19}$$

where

h = head loss, ft of liquid
f = Darcy friction factor, dimensionless
L = Pipe length, ft
D = Pipe internal diameter, ft
V = Average liquid velocity, ft/s
g = Acceleration due to gravity, 32.2 ft/s^2 in English units

Darcy Friction Factor

For laminar flow, with Reynolds number $R < 2000$

$$f = 64/R \tag{3.20}$$

For turbulent flow, with Reynolds number $R > 4000$ (Colebrook-White equation)

$$1/\sqrt{f} = -2\mathrm{Log}_{10}[(e/3.7D) + 2.51/(R\sqrt{f})] \tag{3.21}$$

where

f = Darcy friction factor, dimensionless
D = Pipe internal diameter, in.
e = Absolute pipe roughness, in.

R = Reynolds number of flow, dimensionless

$$f_f = f_d/4 \tag{3.25}$$

where

f_f = Fanning friction factor
f_d = Darcy friction factor

Pressure Drop due to Friction

$$P_m = 0.0605fQ^2(Sg/D^5) \tag{3.27}$$

$$P_m = 0.2421(Q/F)^2(Sg/D^5) \tag{3.28}$$

where

P_m = Pressure drop due to friction, psi per mile
Q = Liquid flow rate, bbl/day
f = Darcy friction factor, dimensionless
F = Transmission factor, dimensionless
Sg = Liquid specific gravity
D = Pipe internal diameter, in.

$$F = 2/\sqrt{f} \tag{3.29}$$

where

F = Transmission factor
f = Darcy friction factor

$$F = -4\text{Log}_{10}[(e/3.7D) + 1.255(F/R)] \quad \text{for turbulent flow } R > 4000 \tag{3.30}$$

In SI units

$$P_{km} = 6.2475 \times 10^{10}fQ^2(Sg/D^5) \tag{3.31}$$

$$P_{km} = 24.99 \times 10^{10}(Q/F)^2(Sg/D^5) \tag{3.32}$$

where

P_{km} = Pressure drop due to friction, kPa/km
Q = Liquid flow rate, m^3/hr
f = Darcy friction factor, dimensionless
F = Transmission factor, dimensionless
Sg = Liquid specific gravity
D = Pipe internal diameter, mm

Colebrook-White Equation

Modified Colebrook-White Equation:

$$F = -4\mathrm{Log}_{10}[(e/3.7D) + 1.4125(F/R)] \quad (3.34)$$

Hazen-Williams Equation

$$h = 4.73L(Q/C)^{1.852}/D^{4.87} \quad (3.35)$$

where

h = Head loss due to friction, ft
L = Length of pipe, ft
D = Internal diameter of pipe, ft
Q = Flow rate, ft^3/s
C = Hazen-Williams coefficient or C-factor, dimensionless

$$Q = 0.1482(C)(D)^{2.63}\,(P_m/Sg)^{0.54} \quad (3.36)$$

where

Q = Flow rate, bbl/day
D = Pipe internal diameter, in.
P_m = Frictional pressure drop, psi/mile
Sg = Liquid specific gravity
C = Hazen-Williams C-factor

$$GPM = 6.7547 \times 10^{-3}(C)(D)^{2.63}(H_L)^{0.54} \quad (3.37)$$

where

GPM = Flow rate, gal/min
H_L = Friction loss, ft per 1000 ft of pipe

In SI units

$$Q = 9.0379 \times 10^{-8}(C)(D)^{2.63}(P_{km}/Sg)^{0.54} \quad (3.38)$$

where

Q = Flow rate, m^3/hr
D = Pipe internal diameter, mm
P_{km} = Frictional pressure drop, kPa/km
Sg = Liquid specific gravity
C = Hazen-Williams C-factor

Shell-MIT Equation

$$R = 92.24(Q)/(D\nu) \quad (3.39)$$

$$Rm = R/(7742) \quad (3.40)$$

where

R = Reynolds number, dimensionless
Rm = Modified Reynolds number, dimensionless
Q = Flow rate, bbl/day
D = Internal diameter, in.
ν = Kinematic viscosity, cSt

$$f = 0.00207/Rm \qquad \text{(laminar flow)} \tag{3.41}$$

$$f = 0.0018 + 0.00662(1/Rm)^{0.355} \qquad \text{(turbulent flow)} \tag{3.42}$$

$$P_m = 0.241(f\,SgQ^2)/D^5 \tag{3.43}$$

where

P_m = Frictional pressure drop, psi/mile
f = Friction factor, dimensionless
Sg = Liquid specific gravity
Q = Flow rate, bbl/day
D = Pipe internal diameter, in.

In SI units

$$P_m = 6.2191 \times 10^{10}(f\,SgQ^2)/D^5 \tag{3.44}$$

where

P_m = Frictional pressure drop, kPa/km
f = Friction factor, dimensionless
Sg = Liquid specific gravity
Q = Flow rate, m^3/hr
D = Pipe internal diameter, mm

Miller Equation

$$Q = 4.06(M)\,(D^5P_m/Sg)^{0.5} \tag{3.45}$$

where

$$M = Log_{10}(D^3SgP_m/cp^2) + 4.35 \tag{3.46}$$

and

Q = Flow rate, bbl/day
D = Pipe internal diameter, in.
P_m = Frictional pressure drop, psi/mile
Sg = Liquid specific gravity
cp = Liquid viscosity, centipoise

In SI Units

$$Q = 3.996 \times 10^{-6}(M)(D^5P_m/Sg)^{0.5} \quad (3.47)$$

where

$$M = \text{Log}_{10}(D^3SgP_m/cp^2) - 0.4965 \quad (3.48)$$

and

Q = Flow rate, m^3/hr
D = Pipe internal diameter, mm
P_m = Frictional pressure drop, kPa/km
Sg = Liquid specific gravity
cp = Liquid viscosity, centipoise

$$P_m = (Q/4.06M)^2(Sg/D^5) \quad (3.49)$$

T. R. Aude Equation

$$P_m = [Q(z^{0.104})(Sg^{0.448})/(0.871(K)(D^{2.656}))]^{1.812} \quad (3.50)$$

where

P_m = Pressure drop due to friction, psi/mile
Q = Flow rate, bbl/hr
D = Pipe internal diameter, in.
Sg = Liquid specific gravity
z = Liquid viscosity, centipoise
K = T. R. Aude K-factor, usually 0.90 to 0.95

In SI Units

$$P_m = 8.888 \times 10^8[Q(z^{0.104})(Sg^{0.448})/(K(D^{2.656}))]^{1.812} \quad (3.51)$$

where

P_m = Frictional pressure drop, kPa/km
Sg = Liquid specific gravity
Q = Flow rate, m^3/hr
D = Pipe internal diameter, mm
z = Liquid viscosity, centipoise
K = T. R. Aude K-factor, usually 0.90 to 0.95

Head Loss in Valves and Fittings

$$h = KV^2/2g \quad (3.52)$$

where

h = Head loss due to valve or fitting, ft

K = Head loss coefficient for the valve or fitting, dimensionless
V = Velocity of liquid through valve or fitting, ft/s
g = Acceleration due to gravity, 32.2 ft/s^2 in English units

Gradual Enlargement

$$h = K(V_1 - V_2)^2/2g \quad (3.53)$$

where V_1 and V_2 are the velocity of the liquid in the smaller-diameter and larger-diameter pipe respectively. Head loss coefficient K depends upon the diameter ratio D_1/D_2 and the different cone angle due to the enlargement.

Sudden Enlargement

$$h = (V_1 - V_2)^2/2g \quad (3.54)$$

Drag Reduction

$$\text{Percentage drag reduction} = 100(DP_0 - DP_1)/DP_0 \quad (3.55)$$

where

DP_0 = Friction drop in pipe segment without DRA, psi
DP_1 = Friction drop in pipe segment with DRA, psi

Explicit Friction Factor Equations

Churchill Equation

This equation proposed by Stuart Churchill for friction factor was reported in *Chemical Engineering* magazine in November 1977. Unlike the Colebrook-White equation, which requires trial-and-error solution, this equation is explicit in f as indicated below.

$$f = [(8/R)^{12} + 1/(A + B)^{3/2}]^{1/12}$$

where

$A = [2.457 \text{ Log}_e(1/((7/R)^{0.9} + (0.27e/D))]^{16}$
$B = (37{,}530/R)^{16}$

The above equation for friction factor appears to correlate well with the Colebrook-White equation.

Swamee-Jain Equation

P. K. Swamee and A. K. Jain presented this equation in 1976 in the *Journal of the Hydraulics Division of ASCE*. It is found to be the best and easiest of all explicit equations for calculating the friction factor.

$$f = 0.25/[\mathrm{Log}_{10}(e/3.7D + 5.74/R^{0.9})]^2$$

It correlates well with the Colebrook-White equation.

Chapter 4

Barlow's Equation for Internal Pressure

$$S_h = PD/2t \tag{4.1}$$

where

S_h = Hoop stress, psi
P = Internal pressure, psi
D = Pipe outside diameter, in.
t = Pipe wall thickness, in.

$$S_a = PD/4t \tag{4.2}$$

where

S_a = Axial (or longitudinal) stress, psi

Internal design pressure in a pipe in English units

$$P = \frac{2T \times S \times E \times F}{D} \tag{4.3}$$

where

P = Internal pipe design pressure, psig
D = Nominal pipe outside diameter, in.
T = Nominal pipe wall thickness, in.
S = Specified minimum yield strength (SMYS) of pipe material, psig
E = Seam joint factor, 1.0 for seamless and submerged arc welded (SAW) pipes (see Table A.11 in Appendix A)
F = Design factor

In SI units

$$P = \frac{2T \times S \times E \times F}{D} \tag{4.4}$$

where

P = Pipe internal design pressure, kPa
D = Nominal pipe outside diameter, mm
T = Nominal pipe wall thickness, mm
S = Specified minimum yield strength (SMYS) of pipe material, kPa

E = Seam joint factor, 1.0 for seamless and submerged arc welded (SAW) pipes (see Table A.11 in Appendix A)
F = Design factor

Line Fill Volume

In English units

$$V_L = 5.129(D)^2 \tag{4.7}$$

where
V_L = Line fill volume of pipe, bbl/mile
D = Pipe internal diameter, in.

In SI units

$$V_L = 7.855 \times 10^{-4} D^2 \tag{4.8}$$

where
V_L = Line fill volume, m^3/km
D = Pipe internal diameter, mm

Chapter 5

Total Pressure Required

$$P_t = P_{friction} + P_{elevation} + P_{del} \tag{5.1}$$

where
P_t = Total pressure required at A
$P_{friction}$ = Total friction pressure drop between A and B
$P_{elevation}$ = Elevation head between A and B
P_{del} = Required delivery pressure at B

Pump Station Discharge Pressure

$$P_d = (P_t + P_s)/2 \tag{5.3}$$

where
P_d = Pump station discharge pressure
P_s = Pump station suction pressure

Equivalent Length of Series Piping

$$L_E/(D_E)^5 = L_A/(D_A)^5 + L_B/(D_B)^5 \tag{5.7}$$

where

L_E = Equivalent length of diameter D_E
L_A, D_A = Length and diameter of pipe A
L_B, D_B = Length and diameter of pipe B

Equivalent Diameter of Parallel Piping

$$Q^2/D_E^5 = Q_{BC}^2/D_{BC}^5 \tag{5.14}$$

$$Q_{BC}^2/D_{BC}^5 = (Q - Q_{BC})^2/D_{BD}^5 \tag{5.15}$$

where

Q = Total flow through both parallel pipes.
Q_{BC} = Flow through pipe branch BC
$(Q - Q_{BC})$ = Flow through pipe branch BD
D_{BC} = Internal diameter of pipe branch BC
D_{BD} = Internal diameter of pipe branch BD
D_E = Equivalent pipe diameter to replace both parallel pipes BC and BD

Horsepower Required for Pumping

$$BHP = QP/(2449E) \tag{5.16}$$

where

Q = Flow rate, bbl/hr
P = Differential pressure, psi
E = Efficiency, expressed as a decimal value less than 1.0

$$BHP = (GPM)(H)(Spgr)/(3960E) \tag{5.17}$$

$$BHP = (GPM)P/(1714E) \tag{5.18}$$

where

GPM = Flow rate, gal/min.
H = Differential head, ft
P = Differential pressure, psi
E = Efficiency, expressed as a decimal value less than 1.0
Spgr = Liquid specific gravity, dimensionless

In SI units

$$\text{Power (kW)} = \frac{Q\,H\,Spgr}{367.46(E)} \tag{5.19}$$

where

Q = Flow rate, m^3/hr

H = Differential head, m
Spgr = Liquid specific gravity
E = Efficiency, expressed as a decimal value less than 1.0

$$\text{Power (kW)} = \frac{Q\,P}{3600(E)} \tag{5.20}$$

where
P = Pressure, kPa
Q = Flow rate, m^3/hr
E = Efficiency, expressed as a decimal value less than 1.0

Chapter 6

Pump Station Discharge Pressure

$$P_D = (P_T - P_S)/N + P_S \tag{6.4}$$

where
P_D = Pump station discharge pressure
P_T = Total pressure required
P_S = Pump station suction pressure
N = Number of pump stations

Line Fill Volume

$$\text{Line fill volume} = 5.129L(D)^2 \tag{6.5}$$

where
D = Pipe internal diameter, in.
L = Pipe length, miles

Chapter 7

Brake Horsepower

$$\text{Pump BHP} = \frac{Q\,H\,Sg}{3960(E)} \tag{7.1}$$

where
Q = Pump flow rate, gal/min
H = Pump head, ft
E = Pump efficiency, as a decimal value less than 1.0
Sg = Liquid specific gravity (for water Sg = 1.0)

In SI units

$$\text{Power kW} = \frac{Q\ H\ Sg}{367.46\ (E)} \tag{7.2}$$

where

Q = Pump flow rate, m^3/hr
H = Pump head, m
E = Pump efficiency, as a decimal value less than 1.0
Sg = Liquid specific gravity (for water $Sg = 1.0$)

Specific Speed

$$N_S = NQ^{1/2}/H^{3/4} \tag{7.3}$$

where

N_S = Pump specific speed
N = Pump impeller speed, RPM
Q = Flow rate or capacity, gal/min
H = Head, ft

Suction Specific Speed

$$N_{SS} = NQ^{1/2}/(NPSH_R)^{3/4} \tag{7.4}$$

where

N_{SS} = Suction specific speed
N = Pump impeller speed, RPM
Q = Flow rate or capacity, gal/min
$NPSH_R$ = NPSH required at the BEP

Impeller Diameter Change

$$Q_2/Q_1 = D_2/D_1 \tag{7.5}$$

$$H_2/H_1 = (D_2/D_1)^2 \tag{7.6}$$

where

Q_1, Q_2 = Initial and final flow rates
H_1, H_2 = Initial and final heads
D_1, D_2 = Initial and final impeller diameters

Impeller Speed Change

$$Q_2/Q_1 = N_2/N_1 \tag{7.7}$$

$$H_2/H_1 = (N_2/N_1)^2 \tag{7.8}$$

where

Q_1, Q_2 = Initial and final flow rates
H_1, H_2 = Initial and final heads
N_1, N_2 = Initial and final impeller speeds

Suction Piping

$$\text{Suction head} = H_S - H_{fs} \quad (7.10)$$

$$\text{Discharge head} = H_D + H_{fd} \quad (7.11)$$

where

H_S = Static suction head
H_D = Static discharge head
H_{fs} = Friction loss in suction piping
H_{fd} = Friction loss in discharge piping

NPSH

$$(P_a - P_v)(2.31/Sg) + H + E1 - E2 - h \quad (7.12)$$

where

P_a = Atmospheric pressure, psi
P_v = Liquid vapor pressure at the flowing temperature, psi
Sg = Liquid specific gravity
H = Tank head, ft
$E1$ = Elevation of tank bottom, ft
$E2$ = Elevation of pump suction, ft
h = Friction loss in suction piping, ft

Chapter 8

Control Pressure and Throttle Pressure

$$P_c = P_s + \Delta P_1 + \Delta P_2 \quad (8.1)$$

where

P_c = Case pressure in pump 2 or upstream pressure at control valve

$$P_{thr} = P_c - P_d \quad (8.2)$$

where

P_{thr} = Control valve throttle pressure
P_d = Pump station discharge pressure

Chapter 9

Thermal Conductivity

$$H = K(A)(dT/dx) \tag{9.1}$$

where

H = Heat flux perpendicular to the surface area, Btu/hr
K = Thermal conductivity, Btu/hr/ft/°F
A = Area of heat flux, ft^2
dx = Thickness of solid, ft
dT = Temperature difference across the solid, °F

In SI units

$$H = K(A)(dT/dx) \tag{9.2}$$

where

H = Heat flux, W
K = Thermal conductivity, W/m/°C
A = Area of heat flux, m^2
dx = Thickness of solid, m
dT = Temperature difference across the solid, °C

Heat Transfer Coefficient

$$H = U(A)(dT) \tag{9.3}$$

where

U = Overall heat transfer coefficient, Btu/hr/ft^2/°F

Other symbols in Equation (9.3) are the same as in Equation (9.1).
In SI units

$$H = U(A)(dT) \tag{9.4}$$

where

U = Overall heat transfer coefficient, W/m^2/°C

Other symbols in Equation (9.4) are the same as in Equation (9.2).

Heat Content Balance

$$H_{in} - \Delta H + H_w = H_{out} \tag{9.6}$$

where

H_{in} = Heat content entering line segment, Btu/hr
ΔH = Heat transferred from line segment to surrounding medium (soil or air), Btu/hr
H_w = Heat content from frictional work, Btu/hr

H_{out} = Heat content leaving line segment, Btu/hr

In SI units Equation (9.6) will be the same, with each term expressed in watts instead of Btu/hr.

Logarithmic Mean Temperature

$$T_m - T_S = \frac{(T_1 - T_S) - (T_2 - T_S)}{\mathrm{Log_e}[(T_1 - T_S)/(T_2 - T_S)]} \qquad (9.7)$$

where

T_m = Logarithmic mean temperature of pipe segment, °F
T_1 = Temperature of liquid entering pipe segment, °F
T_2 = Temperature of liquid leaving pipe segment, °F
T_s = Sink temperature (soil or surrounding medium), °F

In SI units Equation (9.7) will be the same, with all temperatures expressed in°C instead of °F.

Heat Content Entering and Leaving a Pipe Segment

$$H_{in} = w(C_{pi})(T_1) \qquad (9.8)$$

$$H_{out} = w(C_{po})(T_2) \qquad (9.9)$$

where

H_{in} = Heat content of liquid entering pipe segment, Btu/hr
H_{out} = Heat content of liquid leaving pipe segment, Btu/hr
C_{pi} = Specific heat of liquid at inlet, Btu/lb/°F
C_{po} = Specific heat of liquid at outlet, Btu/lb/°F
w = Liquid flow rate, lb/hr
T_1 = Temperature of liquid entering pipe segment, °F
T_2 = Temperature of liquid leaving pipe segment, °F

In SI units

$$H_{in} = w(C_{pi})(T_1) \qquad (9.10)$$

$$H_{out} = w(C_{po})(T_2) \qquad (9.11)$$

where

H_{in} = Heat content of liquid entering pipe segment, J/s (W)
H_{out} = Heat content of liquid leaving pipe segment, J/s (W)
C_{pi} = Specific heat of liquid at inlet, kJ/kg/°C
C_{po} = Specific heat of liquid at outlet, kJ/kg/°C
w = Liquid flow rate, kg/s
T_1 = Temperature of liquid entering pipe segment, °C
T_2 = Temperature of liquid leaving pipe segment, °C

Heat Transfer: Buried Pipeline

$$H_b = 6.28(L)(T_m - T_S)/(Parm1 + Parm2) \quad (9.12)$$

$$Parm1 = (1/K_{ins})Log_e(R_i/R_p) \quad (9.13)$$

$$Parm2 = (1/K_s)Log_e[2S/D + ((2S/D)^2 - 1)^{1/2}] \quad (9.14)$$

where

H_b = Heat transfer, Btu/hr
T_m = Log mean temperature of pipe segment, °F
T_s = Ambient soil temperature, °F
L = Pipe segment length, ft
R_i = Pipe insulation outer radius, ft
R_p = Pipe wall outer radius, ft
K_{ins} = Thermal conductivity of insulation, Btu/hr/ft/°F
K_s = Thermal conductivity of soil, Btu/hr/ft/°F
S = Depth of cover (pipe burial depth) to pipe centerline, ft
D = Pipe outside diameter, ft

In SI units

$$H_b = 6.28(L)(T_m - T_s)/(Parm1 + Parm2) \quad (9.15)$$

$$Parm1 = (1/K_{ins})Log_e(R_i/R_p) \quad (9.16)$$

$$Parm2 = (1/K_s)Log_e[2S/D + ((2S/D)^2 - 1)^{1/2}] \quad (9.17)$$

where

H_b = Heat transfer, W
T_m = Log mean temperature of pipe segment, °C
T_s = Ambient soil temperature, °C
L = Pipe segment length, m
R_i = Pipe insulation outer radius, mm
R_p = Pipe wall outer radius, mm
K_{ins} = Thermal conductivity of insulation, W/m/°C
K_s = Thermal conductivity of soil, W/m/°C
S = Depth of cover (pipe burial depth) to pipe centerline, mm
D = Pipe outside diameter, mm

Heat Transfer: Above-Ground Pipeline

$$H_a = 6.28(L)(T_m - T_a)/(Parm1 + Parm3) \quad (9.18)$$

$$Parm3 = 1.25/[R_i(4.8 + 0.008(T_m - T_a))] \quad (9.19)$$

$$Parm1 = (1/K_{ins})Log_e(R_i/R_p) \quad (9.20)$$

where

H_a = Heat transfer, Btu/hr
T_m = Log mean temperature of pipe segment, °F
T_a = Ambient air temperature, °F
L = Pipe segment length, ft
R_i = Pipe insulation outer radius, ft
R_p = Pipe wall outer radius, ft
K_{ins} = Thermal conductivity of insulation, Btu/hr/ft/°F

In SI units

$$H_a = 6.28(L)(T_m - T_a)/(Parm3 + Parm1) \tag{9.21}$$

$$Parm3 = 1.25/[R_i(4.8 + 0.008(T_m - T_a))] \tag{9.22}$$

$$Parm1 = (1/K_{ins})Log_e(R_i/R_p) \tag{9.23}$$

where

H_a = Heat transfer, W
T_m = Log mean temperature of pipe segment, °C
T_a = Ambient air temperature, °C
L = Pipe segment length, m
R_i = Pipe insulation outer radius, mm
R_p = Pipe wall outer radius, mm
K_{ins} = Thermal conductivity of insulation, W/m/°C

Frictional Heating

$$H_w = 2545(HHP) \tag{9.24}$$

$$HHP = (1.7664 \times 10^{-4})(Q)(Sg)(h_f)(L_m) \tag{9.25}$$

where

H_w = Frictional heat gained, Btu/hr
HHP = Hydraulic horsepower required for pipe friction
Q = Liquid flow rate, bbl/hr
Sg = Liquid specific gravity
h_f = Frictional head loss, ft/mile
L_m = Pipe segment length, miles

In SI units

$$H_w = 1000(Power) \tag{9.26}$$

$$Power = (0.00272)(Q)(Sg)(h_f)(L_m) \tag{9.27}$$

where

H_w = Frictional heat gained, W

Power = Power required for pipe friction, kW
Q = Liquid flow rate, m^3/hr
Sg = Liquid specific gravity
h_f = Frictional head loss, m/km
L_m = Pipe segment length, km

Pipe Segment Outlet Temperature

For buried pipe:

$$T_2 = (1/wC_p)[2545(HHP) - H_b + (wC_p)T_1] \tag{9.28}$$

For above-ground pipe:

$$T_2 = (1/wC_p)[2545(HHP) - H_a + (wC_p)T_1] \tag{9.29}$$

where

H_b = Heat transfer for buried pipe, Btu/hr from Equation (9.12)
H_a = Heat transfer for above-ground pipe, Btu/hr from Equation (9.18)
C_p = Average specific heat of liquid in pipe segment

In SI units

For buried pipe:

$$T_2 = (1/wC_p)[1000(Power) - H_b + (wC_p)T_1] \tag{9.30}$$

For above-ground pipe:

$$T_2 = (1/wC_p)[1000(Power) - H_a + (wC_p)T_1] \tag{9.31}$$

where

H_b = Heat transfer for buried pipe, W
H_a = Heat transfer for above-ground pipe, W
Power = Frictional power defined in Equation (9.27), kW

Chapter 10

Mass Flow

$$Q = AV \tag{10.2}$$

where

A = Cross-sectional area of flow
V = Velocity of flow

Venturi Meter

Velocity in main pipe section:

$$V_1 = C\sqrt{[2g(P_1 - P_2)/\gamma]/[(A_1/A_2)^2 - 1]} \tag{10.6}$$

Velocity in throat:

$$V_2 = C\sqrt{[2g(P_1 - P_2)/\gamma]/[1 - (A_2/A_1)^2]} \tag{10.7}$$

Volume flow rate:

$$Q = CA_1\sqrt{[2g(P_1 - P_2)/\gamma]/[(A_1/A_2)^2 - 1]} \tag{10.8}$$

$$Q = CA_1\sqrt{[2g(P_1 - P_2)/\gamma]/[(1/\beta)^4 - 1]} \tag{10.9}$$

where

P_1 = Pressure in main section of diameter D and area A_1
P_2 = Pressure in throat section of diameter d and area A_2
γ = Specific weight of liquid
C = Coefficient of discharge that depends on Reynolds number and diameters
Beta ratio $\beta = d/D$ and $A_1/A_2 = (D/d)^2$

Flow Nozzle Discharge Coefficient

$$C = 0.9975 - 6.53\sqrt{(\beta/R)} \tag{10.10}$$

where

$$\beta = d/D$$

and R is the Reynolds number based on the main pipe diameter D.

Turbine Meter

$$Q_b = Q_f \times M_f \times F_t \times F_p \tag{10.11}$$

where

Q_b = Flow rate at base conditions, such as 60°F and 14.7 psi
Q_f = Measured flow rate at operating conditions, such as 80°F and 350 psi
M_f = Meter factor for correcting meter reading, based on meter calibration data
F_t = Temperature correction factor for correcting from flowing temperature to the base temperature

F_p = Pressure correction factor for correcting from flowing pressure to the base pressure

Chapter 11

Pressure rise due to sudden closure of a valve:

$$\Delta H = aV/g \tag{11.1}$$

where

a = Velocity of propagation of pressure wave, ft/s
V = Velocity of liquid flow, ft/s
g = Acceleration due to gravity, ft/s^2

Pressure rise due to partial closure of a valve that reduces the velocity from V to V_1:

$$\Delta H = a(V_1 - V)/g \tag{11.2}$$

Wave speed:

$$a = \frac{(K/\rho)^{1/2}}{[1 + C(K/E)(D/t)]^{1/2}} \tag{11.3}$$

where

a = Wave speed, ft/s
K = Bulk modulus of liquid, psi
ρ = Density of liquid, slugs/ft^3
C = Restraint factor, dimensionless
D = Pipe outside diameter, in.
t = Pipe wall thickness, in.
E = Young's modulus of pipe material, psi

In SI units

$$a = \frac{(K/\rho)^{1/2}}{[1 + C(K/E)(D/t)]^{1/2}} \tag{11.4}$$

where

a = Wave speed, m/s
K = Bulk modulus of liquid, kPa
ρ = Density of liquid, kg/m^3
C = Restraint factor, dimensionless
D = Pipe outside diameter, mm

t = Pipe wall thickness, mm
E = Young's modulus of pipe material, kPa

The restraint factor C depends on the type of pipe condition as follows:

Case 1: Pipe is anchored at the upstream end only
Case 2: Pipe is anchored against any axial movements
Case 3: Each pipe section is anchored with expansion joints

Restraint factor C for thin-walled elastic pipes:

$$C = 1 - 0.5\mu \qquad \text{for case 1} \tag{11.5}$$

$$C = 1 - \mu^2 \qquad \text{for case 2} \tag{11.6}$$

$$C = 1.0 \qquad \text{for case 3} \tag{11.7}$$

where μ = Poisson's ratio for pipe material, usually in the range of 0.20 to 0.45 (for steel pipe $\mu = 0.30$).

For thick-walled pipes with D/t ratio less than 25, C values are as follows:

Case 1

$$C = \frac{2t}{D}(1+\mu) + \frac{D(1 - 0.5\mu)}{D + t} \tag{11.8}$$

Case 2

$$C = \frac{2t}{D}(1+\mu) + \frac{D(1-\mu^2)}{D + t} \tag{11.9}$$

Case 3

$$C = \frac{2t}{D}(1+\mu) + \frac{D}{D + t} \tag{11.10}$$

Chapter 12

Pipe Material Cost

$$PMC = 28.1952\ L(D - t)t(Cpt) \tag{12.1}$$

where

PMC = Pipe material cost, $
L = Pipe length, miles
D = Pipe outside diameter, in.

t = Pipe wall thickness, in.
Cpt = Pipe cost, $/ton

In SI units

$$PMC = 0.02463\ L(D - t)t(Cpt) \tag{12.2}$$

where

PMC = Pipe material cost, $
L = Pipe length, km
D = Pipe outside diameter, mm
t = Pipe wall thickness, mm
Cpt = Pipe cost, $/metric ton

References

1. Anon. (1976) *Flow of Fluids through Valves, Fittings and Pipes*. Crane Company.
2. Anon. (1979) *Hydraulic Institute Engineering Data Book*. Hydraulic Institute.
3. Anon. (1981) *Cameron Hydraulic Data*. Ingersoll-Rand.
4. ASCE (1975) *Pipeline Design for Hydrocarbons, Gases and Liquids*. American Society of Civil Engineers.
5. Benedict, R. (1980) *Fundamentals of Pipe Flow*. New York: Wiley.
6. Brater, E. F. and King, H. W. (1982) *Handbook of Hydraulics*. New York: McGraw-Hill.
7. Chaudhry, M. H. (1979) *Applied Hydraulic Transients*. Van Nostrand Reinhold.
8. Cheremisinoff, N. (1979). *Applied Fluid Flow Measurement*. New York: Marcel Dekker.
9. Cheremisinoff, N. (1982) *Fluid Flow*. Ann Arbor, MI: Ann Arbor Science.
10. Lobanoff, V. S. and Ross, R. R. (1985) *Centrifugal Pumps Design and Application*. Houston: Gulf Publishing.
11. Miller, R. W. (1983) *Flow Measurement Engineering Handbook*. New York: McGraw-Hill.
12. Upp, E. L. (1993) *Fluid Flow Measurement*. Gulf Publishing Company.
13. Vennard, J. K. and Street, R. T. (1982) *Elementary Fluid Mechanics*, 6th ed. New York: Wiley.
14. Wylie, E. B. and Streeter, V. L. (1993) *Fluid Transients in Systems*. Prentice Hall.
15. Mott, R. L. (2000) *Applied Fluid Mechanics*, 5th ed. Upper Saddle River, NJ: Prentice Hall.

Index

Milton Keynes UK
Ingram Content Group UK Ltd.
UKHW020023071024
449327UK00032B/2904

9 780367 394158